# A-level Study Guide

# Biology

**Revised and updated for 2003/4**

**Gareth Rowlands**

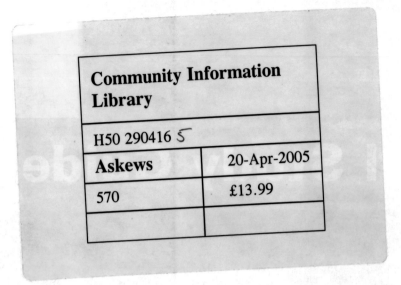
**Gareth Rowlands** is a teacher with over 30 years' experience in secondary schools in England and Wales. His post prior to retirement was as Head of Biology at Ysgol Syr Thomas Jones, Amlwch, Anglesey (1983–98). He also has many years experience as an examiner and is currently an examiner of A-level Biology for the Welsh Joint Education Committee.

**Acknowledgements**

I would like to express my thanks to **Ann Rowlands** for her editorial assistance.

I am indebted to the following examination boards for permission to reproduce questions from their past examination papers:

The Welsh Joint Examining Board
The Associated Examining Board

The examination boards are not responsible for the suggested answers to the questions. The full responsibility for these is accepted by the author.

Series Consultants: Geoff Black and Stuart Wall
Project Manager: Julia Morris

**Pearson Education Limited**

Edinburgh Gate, Harlow
Essex CM20 2JE, England
*and Associated Companies throughout the world*

© Pearson Education Limited 2000, 2003

**British Library Cataloguing in Publication Data**
A catalogue entry for this title is available from the British Library.

ISBN 0-582-78413-1

First published 2000
Reprinted 2001, 2004
Updated 2003

Set by 35 in Univers, Cheltenham
Printed in the UK by Ashford Colour Press

# Molecules and cells

All living organisms are made up of cells, each of which is surrounded by a membrane. These membranes are also found internally where they separate functional units, called organelles, within the cell. One important organelle is the nucleus, which contains the hereditary factors, or genes, of an organism. The manipulation of these genes is called genetic engineering. The structure and properties of biologically significant molecules such as water, carbohydrates, lipids, proteins and nucleic acids are important in understanding metabolic processes such as respiration and protein synthesis, which are controlled by biological catalysts called enzymes.

## Exam themes

The structure and function of organelles
The fluid mosaic model of membrane structure
The properties of biological molecules, recognizing and identifying their general formulae and structure and relating these to their roles
A knowledge of the properties of enzymes and the conditions affecting the way in which they function
The structure and replication of DNA
The role of DNA in protein synthesis
Chromosome behaviour in mitosis
Applications of gene technology
The principles of gene manipulation

## Topic checklist

○ AS  ● A2

| | AQA/A | AQA/B | EDEXCEL | OCR | WJEC |
|---|---|---|---|---|---|
| Molecules 1 | ○ | ○ | ○ | ○ | ○ |
| Molecules 2 | ○ | ○ | ○ | ○ | ○ |
| Nucleic acids | ○ | ○ | ○ | ○ | ○ |
| Replication and protein synthesis | ○ | ○ | ○ | ○ | ○ |
| Enzymes 1 | ○ | ○ | ○ | ○ | ○ |
| Enzymes 2 | ○ | ○ | ○ | ○ | ○ |
| Cell structure and cell membrane | ○ | ○ | ○ | ○ | ○ |
| Cell organelles | ○ | ○ | ○ | ○ | ○ |
| Chromosome structure and mitosis | ○● | ○ | ○ | ○ | ○ |
| Gene technology | ○ | ○ | ○● | ○● | ○ |

# Molecules 1

Even though organisms are incredibly complicated they are all made up of the same type of molecules. To understand how a molecule carries out a particular function in an organism you will need to look at the properties of the molecule. The molecule gets its properties from its structure so you must look at the structural level to understand the processes of life.

## Water

Water is vital to all living organisms, as a constituent of cells, a reactant in metabolic reactions, a solvent and, for aquatic organisms, a habitat.

### Properties of water molecules

→ Polar, with negative charge on hydroxyl ions and positive charge on hydrogen ions. The attraction of oppositely charged poles of water molecules causes them to group together. Attractive forces form hydrogen bonds.
→ High specific heat, which is important:
  → inside cells, where metabolic reactions are enzyme controlled.
  → externally to provide a constant environment for aquatic organisms.
→ High latent heat, which is significant in temperature control where heat is used for vaporization of water when sweating.
→ Maximum density at 4 °C.
→ High surface tension.

## Carbohydrates

These are organic molecules containing carbon, hydrogen and oxygen. Carbohydrates are of three types: monosaccharides, disaccharides and polysaccharides. They have two main functions:

→ an easily accessible source of energy in both plants and animals
→ cellulose has a structural role in plant cell walls

### Monosaccharides

These have the general formula $(CH_2O)_n$ where $n$ is any number from 3 to 9.

$n = 3$: trioses, important as intermediates in metabolism
$n = 5$: pentoses, used in the formation of nucleic acids
$n = 6$: hexoses, e.g. glucose, the main source of energy

All sugars share the formula $C_6H_{12}O_6$ but they differ in their molecular structure, i.e. they are isomers. Glucose exists as two isomers, the α form and the β form. These different forms result in considerable biological differences when they form polymers such as starch and cellulose.

**Examiner's secrets**

You are not expected to reproduce molecular structures in exams. The questions generally provide the structure of molecules and require you to show how two molecules bond together.

**The jargon**

*Polar* molecules are those with unevenly distributed charges.

*Hydrogen bonds* are individually weak but collectively strong.

A high *specific heat* is where a large amount of heat energy is needed to cause a small rise in temperature.

A high *latent heat* is where a great deal of heat energy is needed to change from liquid state to vapour.

**Examiner's secrets**

Numbering carbon atoms from 1 to 6 in the glucose molecule is useful.

**Action point**

Draw diagrams of the two forms of glucose.

**Checkpoint 1**

How many carbons are there in a pentose sugar?

**Checkpoint 2**

How do α and β forms of glucose differ?

## Disaccharides

Consist of two monosaccharide units linked together by the formation of a glycosidic bond *with the elimination of water*. This is a condensation reaction.

## Polysaccharides

These are formed from very large numbers of monosaccharide units linked together. The two main polysaccharides to consider are:

→ starch (in plants) and glycogen (in animals)
  → they are built up of α glucose molecules
  → they act as energy stores
  → their properties include being insoluble and compact (because chains may be folded), do not alter osmotic concentrations, and do not easily diffuse out of the cell, making them ideal for storage
→ cellulose
  → consists of long chains of β glucose molecules cross-linked to each other by hydrogen bonds and these chains are grouped together into microfibrils
  → has a structural role as it gives a high tensile strength to plant cell walls

Table showing classification of selected carbohydrates.

| Monosaccharides | Disaccharides | Polysaccharides |
|---|---|---|
| glucose | maltose | starch |
| fructose | sucrose | glycogen |
| galactose | lactose | cellulose |

**Exam questions**    answers: page 24

1  (a) Cellulose is the most abundant naturally occurring compound on Earth. State precisely where cellulose is most likely to be found.

   (b) The diagram below shows part of a molecule of cellulose.

   (i)   Name molecule Z.
   (ii)  Ring the term which correctly describes molecule Z.

        hexose   pentose   tetrose   triose

   (iii) Describe how units in the diagram are arranged in a complete molecule of cellulose.
   (iv)  Name one property this molecular structure gives to cellulose.

(8 min)

---

**Action point**

Draw a diagram showing the formation of a disaccharide from two monosaccharides.

**Checkpoint 3**

When two monosaccharides combine a molecule of water is released. Is this a hydrolysis or condensation reaction?

**Action point**

Make notes and draw diagrams of the structures of the different types of carbohydrates.

**Test yourself**

Make sure you can relate the structure of polysaccharides to their functions.

**Checkpoint 4**

Is cellulose made up of polymers of α or β glucose?

**Action point**

List the disaccharides and polysaccharides in the table and briefly describe where they are found in plants and/or animals.

**Examiner's secrets**

You are rarely expected to memorize structural diagrams of polysaccharides.

5

# Molecules 2

You need to study the structure of fats and proteins, both of which have important roles. Fats are particularly important in the structure of membranes. Proteins form the structural basis of all living cells.

## Fats (lipids)

Lipids are **triglycerides**, composed of carbon, hydrogen and oxygen, but their oxygen content is very low. They are formed by condensation reactions between glycerol and molecules of fatty acids. They are non-polar and so are insoluble in water. Different triglyceride fats are formed from different fatty acids.

Their functions include:

→ acting as an energy store
→ acting as a heat insulator
→ protecting delicate internal organs

### Phospholipids

Phospholipids are lipids in which:

→ one of the fatty acid groups is replaced by a phosphate group
→ the phosphate group is hydrophilic (polar)
→ the rest of the molecule is hydrophobic (non-polar)

They are important in the formation and functioning of membranes in cells.

## Proteins

→ Proteins are built up from linear sequences of **amino acids**. About 20 different amino acids are used to make up proteins.
→ They differ from carbohydrates and lipids in that in addition to carbon, hydrogen and oxygen they always contain nitrogen. Many proteins also contain sulphur and sometimes phosphorus.
→ They possess an **amino** group ($NH_2$) at one end of the molecule, and a **carboxyl** group (COOH) at the other end.

### Properties of proteins
Proteins are:

→ crystalline and colourless
→ amphoteric and so can act as buffers

### Structure of proteins
Proteins are built up from a linear sequence of amino acids. The amino group of one amino acid reacts with the carboxyl group of another with the elimination of water (condensation reaction). The bond that is formed is called a **peptide** bond and the resulting compound is a **dipeptide**. A number of amino acids joined in this way is called a **polypeptide**.

### Check the net

You'll find up-to-date information about proteins at www.indiana.edu/~cheminfo/09-16.html

### Action point

Draw a diagram to show the formation of a triglyceride from one molecule of glycerol and three molecules of fatty acid.

### Checkpoint 1

A lipid is insoluble in water. How does the addition of a phosphate group affect this property?

### The jargon

*Hydrophilic* means to attract water.
*Hydrophobic* means to repel water.

### Links

Look at the structure of the cell membrane on pages 16 and 17.

### Checkpoint 2

Why are the molecules of an amino acid described as polar?

### Test yourself

Write down the generalized structural formula of an amino acid.

### The jargon

An *amphoteric* substance can act as both an acid and a base.
*Buffers* resist changes in pH.

### Action point

Draw a diagram to show how two amino acids join together to form a dipeptide.

Four levels of protein structure exist.

1 The **primary** structure of a protein is the sequence of amino acids in its polypeptide chain. The proteins differ from each other in the variety, numbers and orders of their constituent amino acids.

2 The **secondary** structure is the shape that the polypeptide chain forms as a result of hydrogen bonding. This is most often a spiral known as the α helix. An alternative is a pleated sheet occurring as a flat zig-zag chain.

3 The **tertiary** structure is formed by the bending and twisting of the polypeptide helix into a compact three-dimensional (3D) structure. The shape is maintained by disulphide, ionic and hydrogen bonds.

4 The **quaternary** structure arises from a combination of a number of different polypeptide chains. These are associated with non-protein groups and form large, complex molecules, e.g. a haemoglobin molecule is formed from four separate polypeptide chains, each associated with a complex iron-containing prosthetic group, called haem.

## Classification of proteins

Proteins can be classified according to their structure:

→ **fibrous** proteins (perform structural functions) are
  → insoluble in water
  → strong and tough
  → made up of long polypeptide chains or sheets often with numerous cross-linkages
→ **globular** proteins (function as enzymes, antibodies, plasma proteins and hormones) are
  → folded as globular complex 3D molecules
  → soluble in water

**Action point**

Draw sketches of the different levels of protein structure.

**Checkpoint 3**

How is the secondary structure of a protein maintained?

**Checkpoint 4**

How are ionic and hydrogen bonds affected by heat?

**Checkpoint 5**

What type of proteins are enzymes?

**Action point**

Make a list of the uses of proteins. You'll be surprised how many uses there are!

---

**Exam questions**                                                answers: page 24

1 The diagram below shows the structural formula of a typical amino acid.

  (a) (i)  Name the parts of the molecule labelled A and B.

$$A \qquad \overset{H}{\underset{R}{|}} \qquad B$$

$$H_2N \; + \; C \; + \; COOH$$

$$\underset{H}{|}$$

   (ii)  State what R would represent in the simplest amino acid.

  (b) (i)  Draw a diagram to show two molecules of amino acids joined by means of a peptide bond.

   (ii)  Label the peptide bond on your diagram.

   (iii) Name the type of reaction involved when two amino acids combine in this way.

                                                                    (8 min)

**Don't forget**

Include the biochemical tests for a reducing sugar, a non-reducing sugar, starch, protein and lipids.

# Nucleic acids

Nucleic acids are the molecules that contain genetic information and are found in all living cells and viruses. Two types of nucleic acid are found in cells: deoxyribonucleic acid (DNA) and ribonucleic acid (RNA). The structure of DNA was worked out by Watson and Crick in the 1950s.

**Check the net**

You'll find multiple choice questions on nucleic acids at www.biology.arizona.edu

## Structure of nucleic acids

Nucleic acids are built up of units called **nucleotides** (see diagram below).

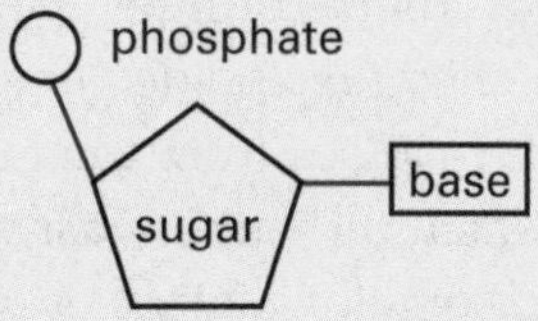

Individual nucleotides are made up of three parts which combine by condensation reactions. These are:

- phosphoric acid (phosphate $H_3PO_4$)
- pentose sugar, of which there are two types
  - in RNA the sugar is ribose
  - in DNA the sugar is deoxyribose
- organic base, of which there are five, divided into two groups
  - the **pyrimidine** bases are thymine, cytosine and uracil
  - the **purine** bases are adenine and guanine

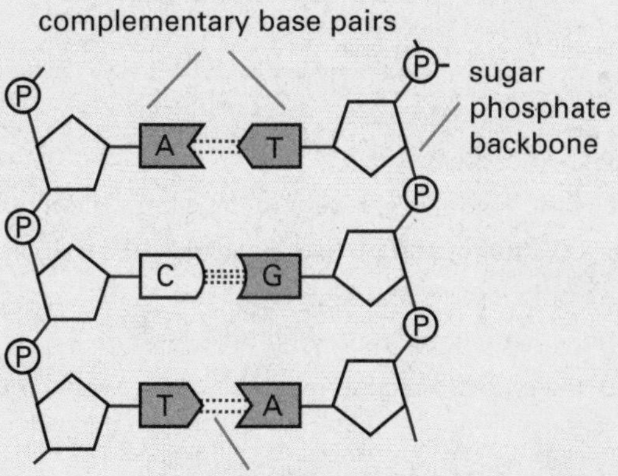

**The jargon**

*Nitrogenous bases* derived from *pyrimidines* are single ring structures whereas those derived from *purines* are double ring structures.

## DNA

The structure is as follows:

- it is a double-stranded polymer of nucleotides (polynucleotide)
- each polynucleotide may contain many million nucleotide units
- it is in the form of a **double helix**, the shape of which is maintained by hydrogen bonding
- the pentose sugar is always deoxyribose
- it contains four organic bases, adenine, guanine, cytosine and thymine
- each strand is linked to the other by pairs of organic bases
- cytosine always pairs with guanine, adenine always pairs with thymine, and the bases are joined by hydrogen bonds

**The jargon**

A *polymer* is a large number of repeating identical units.

**Speed learning**

Complementary base pairs are A–T G–C.

**Checkpoint 1**

Write out the complementary sequence of bases to AGCCTACGT.

This diagram shows how nucleotides are arranged relative to each other in part of a DNA chain.

## RNA

→ RNA is a single-stranded polymer of nucleotide.
→ It contains the pentose sugar, ribose.
→ It contains the organic bases adenine, guanine, cytosine and uracil.

There are three types of RNA.

→ *Ribosomal RNA (rRNA)* is found in the cytoplasm and is a large complex molecule made up of both double and single helices.
→ *transfer RNA (tRNA)* is a small single-stranded molecule. It forms a clover-leaf shape, with one end of the chain ending in a cytosine–cytosine–adenine sequence where the amino acid it carries attaches itself. At the opposite end of the chain is a sequence of three bases called the anticodon (see diagram below).
→ *Messenger RNA (mRNA)* is a long single-stranded molecule, formed into a helix. It is manufactured in the nucleus and passes into the cytoplasm where it associates with the ribosomes.

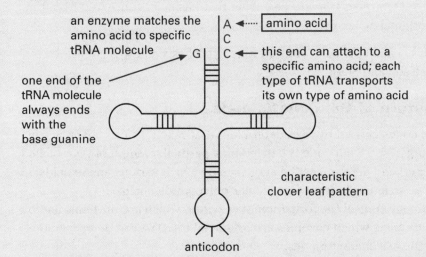

**Action point**

Make your own notes and diagrams on the three different types of RNA.

**Links**

Protein synthesis is found on pages 10 and 11.

**Watch out!**

In RNA uracil replaces thymine, so A pairs with U.

**Examiner's secrets**

You can expect to have a question on some aspect of DNA structure.

**The jargon**

The *anticodon* lines up alongside the appropriate *codon* on the mRNA during protein synthesis.

**Checkpoint 2**

If DNA is likened to a 'ladder' with alternating phosphate and deoxyribose molecules making up the 'uprights', what are the 'rungs' made up of?

**Checkpoint 3**

State two ways in which the structure of RNA differs from that of DNA.

**Checkpoint 4**

Which of RNA and DNA remains constant for all cells (except gametes) of a species?

**Test yourself**

Draw up a table of differences between RNA and DNA.

**Exam questions**                                          answers: page 24

1   The diagram below represents a small portion of DNA.

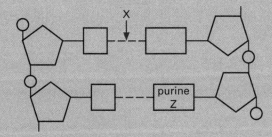

(a) Name:
   (i)   the bond labelled X
   (ii)  one base which could be represented by Z
   (iii) two elements other than carbon, hydrogen and oxygen which are present in DNA
(b) Describe three ways in which the molecular structure of DNA differs from mRNA.

(9 min)

# Replication and protein synthesis

DNA carries the coded description or blueprint of the organism and not only determines which reactions take place but also ensures that this information is transmitted to the next generation. You need to study the two roles that DNA plays, in replication and in protein synthesis.

**Check the net**

You'll find multiple choice questions on protein synthesis at www.biology.arizona.edu

**Links**

Look at the section on mitosis on pages 20 and 21.

**The jargon**

*Complementary bases* are the bases that pair with each other, e.g. A–T, G–C.

**Examiner's secrets**

You may be asked to provide supporting evidence for a theory.

## Replication

When cells divide, each daughter cell must receive an exact copy of the genetic material. It is the DNA molecule that replicates or makes copies of itself. It does this as follows.

➜ DNA unwinds and as the strands separate DNA polymerase catalyses the addition of free nucleotides to the exposed bases.
➜ Each chain acts as a template so that free nucleotides can be joined to their complementary bases by DNA polymerase.
➜ The final result is two DNA molecules each made up of one newly synthesized chain and one chain that has been conserved from the original molecule.

This is called the semi-conservative hypothesis and was confirmed by Meselson and Stahl.

## Nature of the genetic code

The codes carried by DNA determine what reactions can take place in an organism. Each code is carried on a particular length of DNA (called a **gene**) and determines the sequence by which various amino acids are joined together to form a particular polypeptide chain.

Genes control the formation of enzymes, which are proteins, and by determining which enzymes are produced the DNA can determine the organism's characteristics.

➜ It is the sequence of bases in the DNA chain that codes for the sequence of amino acids in a polypeptide.
➜ Each amino acid is coded for by three bases (the triplet code), called a **codon**. Since there are four bases, the number of different codons that are possible is 64, more than enough to code for 20 different amino acids.
➜ All the codons are universal, i.e. they are exactly the same for all organisms.
➜ The code is non-overlapping, in that each triplet is read separately.

**Checkpoint 1**

If the number of bases in a gene coding for a polypeptide is 672, estimate the number of amino acids in the polypeptide chain.

**Links**

Gene mutation on page 148.

**Watch out!**

Don't confuse replication and protein synthesis.

## Protein synthesis

There are four main stages in the formation of a protein:

➜ the synthesis of amino acids
➜ transcription (formation of mRNA)
➜ amino acid activation
➜ translation

DNA in the nucleus acts as a template for the production of mRNA, which conveys the instructions needed for protein synthesis from the nucleus to the cytoplasm. The function of the ribosomes is to provide a suitable surface for the attachment of mRNA and the assembly of protein.

The process of protein synthesis occurs as follows.

1  A very short section of the DNA molecule carries the code for the synthesis of a polypeptide chain. The double-stranded DNA first untwists and then unzips in the relevant region.
2  Free RNA nucleotides then align themselves opposite one of the two strands. Because of the complementary relationship between the bases in DNA and the free nucleotides, cytosine in the DNA attracts a guanine, guanine a cytosine, thymine an adenine, and adenine a uracil.
3  RNA polymerase moves along the DNA adding one complementary RNA nucleotide at a time, resulting in the synthesis of a molecule of mRNA alongside the unzipped portion of DNA. This process is called transcription. The mRNA moves out through the nuclear pore and attaches itself to a ribosome.
4  Activation is the process by which amino acids combine with tRNA using energy from ATP. Each type of tRNA binds with a specific amino acid. The tRNA molecules with attached amino acids now move towards the ribosome.
5  The ribosome acts as a framework moving along the mRNA, reading the code and holding the codon–anticodon complex together until two amino acids join (condensation). The ribosome moves along adding one amino acid at a time until the polypeptide chain is assembled (see diagram below).
6  This process is called translation, whereby a specific sequence of amino acids is formed in accordance with the codons on the mRNA.
7  A group of ribosomes moving along one after the other is called a polysome system. Each time one ribosome moves along the mRNA a molecule of protein is produced. Polypeptides are then assembled into proteins.

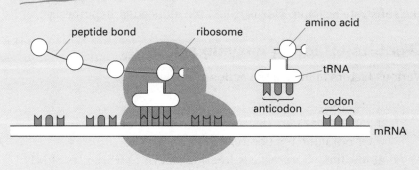

**Action point**

Draw simple diagrams of the various stages in the process of protein synthesis.

**Checkpoint 2**

In which organelle does protein synthesis occur?

**Checkpoint 3**

Name the molecule that moves through the nuclear pore and attaches to a ribosome.

**The jargon**

Each tRNA differs in the composition of a triplet of bases called the *anticodon*.

**Checkpoint 4**

Which molecule carries an amino acid to a ribosome?

**Checkpoint 5**

Name the molecule which is transcribed, but not translated.

**Checkpoint 6**

Which molecule is the site of the anticodon?

**Checkpoint 7**

Name the molecule that is the site of the codon.

**Examiner's secrets**

There is a good chance that there will be a question on protein synthesis. You may be asked a question about a specific part of the process, e.g. to distinguish between transcription and translocation; to decode base sequences; or to write an essay on the complete process.

**Exam question**                                  answer: page 24

Describe the functions of the various forms of nucleic acids in protein synthesis. (15 min)

# Enzymes 1

Metabolic reactions occur quickly and thousands of reactions are taking place at the same time in cells. Order and control is essential if reactions are not to interfere with each other. These features of metabolism are made possible by the action of enzymes. For centuries humans have used enzymes to produce bread, cheese, yoghurt, wine and beer. But now they are used on a wide commercial scale in the pharmaceutical and agrochemical industries and for analytical purposes.

## Properties of enzymes ●●●

**Enzymes** have the following properties:

- → are large protein molecules
- → act as catalysts
- → lower activation energy
- → are reversible in their action
- → are very efficient, having high turnover numbers
- → can be denatured by heat
- → most are specific

## How enzymes work ●●●

Enzymes react with another molecule called a **substrate**. Each enzyme has its own special shape, with an area (the **active site**) on to which the substrate molecules bind (see diagram below).

  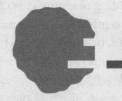

enzyme + substrate     enzyme substrate complex     enzyme + products

Modern interpretations of the lock and key theory suggest that in the presence of the substrate the active site may change in order to select the substrate's shape. This is called the induced fit hypothesis.

## Factors affecting enzyme action ●●●

Various factors influence the action of enzymes:

- → temperature
  - → many enzymes are denatured by temperatures above 45 °C
  - → their optimum temperature is about 37 °C
  - → at low temperatures, e.g. freezing, enzymes work very slowly
- → pH – individual enzymes have a narrow optimum range
- → substrate concentration – the Michaelis constant is the concentration of substrate needed to make the reaction proceed at half its maximum rate. A low Michaelis constant means the reaction proceeds rapidly because the enzyme and substrate have a high affinity for each other.

### Links

Digestive enzymes on page 30. There is a question on 'immobilized enzymes' in the Resources section.

### Checkpoint 1

Why is catalase called the fastest enzyme in the west?

### Action point

Draw a graph to illustrate how enzymes reduce the level of activation energy in a chemical reaction.

### Checkpoint 2

What is meant by the term 'active site'?

### Test yourself

With the aid of diagrams, explain the 'lock and key' theory of enzyme action.

### Checkpoint 3

How does temperature affect the rate of molecular collisions?

### Checkpoint 4

If all the active sites of the enzyme are occupied what is the effect of increasing the substrate concentration?

### Action point

Draw graphs to illustrate these factors.

### The jargon

*Affinity* means chemical attraction.

### The jargon

Immobilized enzymes are trapped and held in a framework of cellulose which is permeable to the substrate and products of the reaction.

## Enzyme cofactors

A cofactor is a non-protein substance that is essential for some enzymes to function efficiently. Cofactors may be metallic ions, or low molecular weight organic molecules that are either temporarily or permanently attached to the enzyme molecule. They are of three types: **activators**, **coenzymes** and **prosthetic groups**.

## Uses of enzymes ●●●

Enzymes are used in:

→ the food industry, e.g. pectins in plant cell walls. When fruits are mashed, pectins cause changes, including colour and flavour deterioration. When fruit ripen, pectinases are produced causing the fruit to soften. Fruit juice manufacturers add industrial pectinase between mashing and pressing to break down pectin so that a good quality, clear juice is obtained.

→ the chemical industry, e.g. proteases are used in biological detergents.

→ in medicine, e.g. in biosensors. The specificity and sensitivity of enzymes is made use of in enzyme-based devices. Biosensors can be used for the rapid and accurate detection of minute traces of biologically important molecules:
  → in pregnancy testing kits
  → monitoring blood sugar levels in diabetics

The enzyme is immobilized on to the surface of an electrode by being enclosed in small alginate beads. It may be used to measure quantities of urea in blood or urine samples (see diagram below). The electrode can detect changes in substrate or product, temperature changes or optical properties.

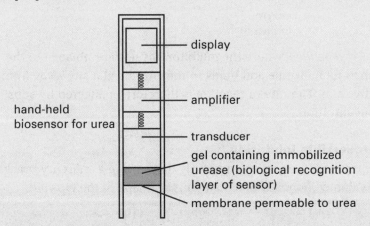

hand-held biosensor for urea

- display
- amplifier
- transducer
- gel containing immobilized urease (biological recognition layer of sensor)
- membrane permeable to urea

Immobilized enzymes are also used widely in industrial processes, such as fermentation, as they can readily be recovered for reuse. Biosensors are also used in fermenters to provide a rapid, sensitive and specific measurement of products.

**Action point**

Write your own notes on other uses of enzymes.

**The jargon**

*Biosensor* describes the association of a *biomolecule*, such as an enzyme, with a *transducer*, which produces an electrical signal in response to substrate transformation.

**Action point**

Write brief notes on additional methods of immobilizing enzymes.

**Checkpoint 5**

State one advantage of immobilized enzymes other than making them easier to reuse than 'free' enzymes.

**Checkpoint 6**

How do enzymes differ from industrial inorganic catalysts?

**Exam question**                                    answer: page 25

Explain fully why almost all biological reactions are catalysed by enzymes and show how the structure of enzymes is adapted to this role. (10 min)

**Links**

Fermentation is dealt with in detail on pages 116 and 117.

# Enzymes 2

The actions of enzymes may be enhanced or inhibited by various substances, some found in the cell and others that are absorbed from the external environment. Certain drugs and pesticides alter metabolism by enzyme inhibition.

## Enzymes and inhibitors

The rate of enzyme-controlled reactions may be decreased by the presence of inhibitors. There are two main types of inhibition, **reversible** and **non-reversible**.

### Reversible inhibition

The effect of this type of inhibitor is temporary. There is no permanent damage to the enzyme because the association of the inhibitor with the enzyme is a loose one. Removal of the inhibitor allows the enzyme to function normally. There are two types of reversible inhibitors.

→ *Competitive*, where the inhibitor is structurally similar to the substrate. The more substrate molecules the greater the chance of finding active sites, leaving fewer to be occupied by the inhibitor (see diagram below).

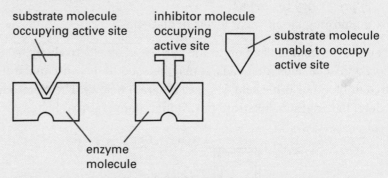

→ *Non-competitive*, where the inhibitor has no resemblance to the substrate molecule and binds to the enzyme at a site away from the active site. The rate of reaction is therefore unaffected by substrate concentration.

### Non-reversible inhibition

These leave the enzyme permanently damaged, e.g. mercury breaks the disulphide bonds which maintain the shape of the enzyme molecule.

### End-product inhibition

In some metabolic pathways the end-product of the pathway may act as an inhibitor. This end-product inhibition is an example of a negative feedback mechanism, preventing the unnecessary accumulation of a metabolite.

**Watch out!**

Don't confuse the different types of inhibitors.

**Checkpoint 1**

If a mixture is prepared containing an enzyme, a competitive inhibitor and a small quantity of substrate, what would be the probable effect on the rate of reaction if more substrate is added?

**Test yourself**

Draw a similar diagram to show how a non-competitive inhibitor affects enzyme action.

**Checkpoint 2**

In terms of attachment to the enzyme what is the difference between competitive and non-competitive inhibitors?

**Checkpoint 3**

What type of inhibition is affected by the concentration of the substrate?

**Checkpoint 4**

In terms of definition what is the main difference between reversible and non-reversible inhibition?

**The jargon**

*Metabolites* are the molecules involved in metabolic reactions.

## Allosteric effectors

These are substances present in cells that reversibly bind with the enzyme away from the active site yet cause a change in the structure of the active site. They are of two types:

→ allosteric activators, which speed up catalysis
→ allosteric inhibitors, which slow down the reaction and provide a way of controlling enzyme activity in metabolism

---

**Exam questions** answers: page 25

1 The graph below shows the relationship between substrate concentration and the initial rate of an enzyme-catalyzed reaction under different conditions.

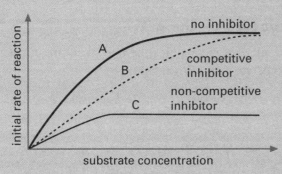

(a) Explain:
   (i) the shape of curve A
   (ii) the difference between the shapes of curve B and curve C
(b) (i) Explain what is meant by the induced fit model of enzyme action.
   (ii) Suggest how this may provide a better explanation for the effects of a non-competitive inhibitor than the lock and key model. (8 min)

2 The thermostability of enzymes is important in industrial use. One such enzyme is papain. The table below compares the thermostability of the enzyme papain in soluble and in immobilized forms.

| Temperature (°C) | Rate of reaction compared with rate at 25 °C (%) | |
| --- | --- | --- |
| | Soluble ('free') | Immobilized |
| 30 | 100 | 100 |
| 40 | 100 | 100 |
| 50 | 90 | 100 |
| 60 | 78 | 90 |
| 70 | 38 | 78 |
| 80 | 10 | 50 |

(a) Explain in terms of protein structure why enzymes are inactivated by high temperatures.
(b) Using the table give evidence that the thermostability of papain is increased by immobilizing it.
(c) Briefly describe one method which might have been used to immobilize the enzyme. (6 min)

# Cell structure and cell membrane

The cell is the basic structural and functional unit of an organism. The distinction between eukaryotic and prokaryotic cells is one of the most fundamental dividing lines between living organisms. The effective boundary between the cell and its environment is the cell membrane.

## Prokaryotic and eukaryotic cells

*Prokaryotic* cells were probably the first forms of life on Earth. This type of cell is found in bacteria and blue-green algae. Their DNA is not enclosed within a nuclear membrane and there are no membrane-bounded organelles.

*Eukaryotic* cells are typical of the great majority of organisms including all animals and plants. They possess membrane-bound organelles such as mitochondria and chloroplasts, and a distinct membrane-bound nucleus.

## Ultrastructure of the cell

The electron microscope enables the fine structure of cells and their organelles to be seen, as shown below.

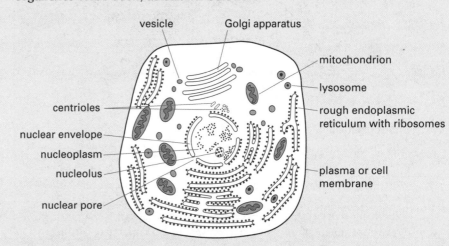

## Cell membrane

The cell membrane is made up almost entirely of proteins and phospholipids. The fluid mosaic model was suggested by Singer and Nicholson (see diagram opposite). They proposed that:

→ there is a bimolecular phospholipid layer with inwardly directed hydrophobic lipid tails and outwardly directed hydrophilic phosphate heads
→ associated with the bilayer is a variety of protein molecules
    → some of the proteins occur on or in only one of the layers (extrinsic proteins)
    → some proteins extend across both layers (intrinsic proteins)
→ the phospholipid layer can move, i.e. it is fluid and in surface view the proteins are dotted throughout the layer in a mosaic arrangement

**Checkpoint 1**

Construct a table of differences between eukaryote and prokaryote cells.

**Checkpoint 2**

What term is applied to the outer layer of an animal cell?

**Checkpoint 3**

What additional outer layer is found in plant cells only?

**Test yourself**

Explain in your own words why Singer and Nicholson's theory is aptly named.

**Links**

Cell organelles are described on pages 18 and 19, and phospholipids are described on page 6.

**The jargon**

*Bimolecular* in this context means phospholipid molecules stretched out to form two layers.

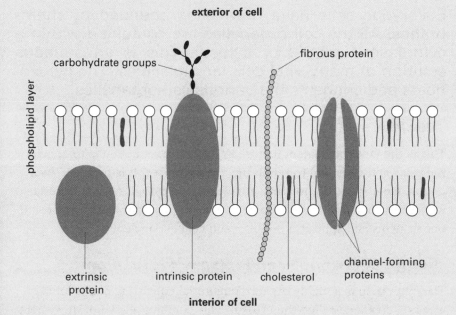

exterior of cell

carbohydrate groups

fibrous protein

phospholipid layer

extrinsic protein

intrinsic protein

cholesterol

channel-forming proteins

interior of cell

**Checkpoint 4**

What is the difference between extrinsic and intrinsic proteins found in cell membranes?

**Checkpoint 5**

Why is the membrane described by the fluid mosaic model referred to as a 'dynamic structure'?

**Checkpoint 6**

What type of molecules freely cross the cell membrane?

**Checkpoint 7**

How does water cross the membrane barrier?

**Links**

Facilitated diffusion and active transport are dealt with in detail on pages 28 and 29 respectively.

**Links**

Look back at the structure of cellulose on page 5.

**Action point**

Make a table to show the differences between plant and animal cells.

## The membrane as a barrier

→ Small uncharged molecules, such as oxygen and carbon dioxide, freely cross as they are soluble in the lipid part of the membrane.

→ Fatty acid tails stop water molecules and water-soluble molecules from moving across.

→ Charged particles (ions) and relatively large molecules such as glucose do not readily pass through the membrane because they are relatively insoluble in lipid. Certain proteins assist such particles to pass in or out of the cell. These proteins are of two types:

\ → channel proteins

ᴕ→ carrier proteins

## Cellulose cell wall

The membrane of plant cells is bounded by an additional layer, the cell wall, consisting of cellulose microfibrils embedded in a polysaccharide matrix. This is a tough but slightly elastic layer which protects and maintains the shape of the cell. Whereas the membrane is selectively permeable, the cell wall is fully permeable.

**Exam questions**                                   answers: pages 25–6

1  With reference to the cell membrane explain fully what is meant by each of these two terms:
   (a) bilayer structure
   (b) fluid mosaic

                                                     (15 min)

# Cell organelles

**Action point**

Draw up a table relating structure to function for the various organelles.

**Checkpoint 1**

In the context of protein synthesis what is the most likely role of the nuclear pores?

Eukaryotic cells have organelles bounded by membranes. All the cell organelles are contained within a cytoplasmic matrix or cytosol. This is an aqueous solution of many vital cellular molecules. The cytosol holds and connects the various cell organelles.

## Nucleus

This is the most prominent feature in the cell. Its function is to control the cell's activities and to retain the chromosomes. It is bounded by a double membrane, the **nuclear envelope**. The cytoplasm-like material within the nucleus is called the nucleoplasm. Within the nucleus are one or two small spherical bodies, each called a nucleolus.

## Rough and smooth endoplasmic reticulum

The cytoplasm is a highly organized material consisting of a soluble ground substance called the **cytosol**. This contains an elaborate system of parallel, flattened cavities lined with a thin membrane. This system is known as the **endoplasmic reticulum** (ER). The cavities are interconnected and the lining membranes are continuous with the nuclear envelope. There are two types of ER:

**Checkpoint 2**

Which organelle has a double membrane and is only found in plant cells?

→ *rough ER*, where the membranes are lined with ribosomes, the sites of protein synthesis
→ *smooth ER*, where the membranes lack ribosomes, being concerned with the synthesis and transport of lipids

## Golgi apparatus

This is similar in structure to ER but is more compact. It receives, sorts, modifies and delivers proteins and lipids.

**Links**

You will need to study chloroplasts and mitochondria in detail in order to understand the biochemistry of photosynthesis and respiration described in the section on page 69.

## Chloroplast

The chloroplast is found in plant cells only. It is bounded by a double membrane. Within the chloroplast are two regions:

→ the **stroma**, which is a colourless gelatinous matrix in which the grana are embedded
→ the **grana**, each made up of a number of closed flattened sacs called **thylakoids**, within which are found the photosynthetic pigments such as chlorophyll

**Check the net**

You'll find up-to-date information on the mitochondrion at cellbio.utmb.edu/cellbio/mitoch1.htm

## Mitochondrion

**Checkpoint 3**

Which organelle is concerned with respiration?

The mitochondrion is bounded by two thin membranes separated by a narrow fluid-filled space. The inner membrane is folded inwards to form extensions called **cristae**. The interior contains an organic matrix containing numerous chemical compounds. The chemical reactions of aerobic respiration take place in the mitochondrion. Some of the reactions take place in the matrix, while others occur on the inner membrane and in the cytosol. The cristae increase the surface area on which the respiratory processes take place.

## Lysosomes

The lysosomes contain and isolate digestive enzymes from the remainder of the cell. Digestion is carried out in a membrane-lined vacuole into which several lysosomes may discharge their contents. One of their functions is to destroy worn out organelles in the cell.

## Ribosomes

These are made up of ribosomal RNA and protein. They are important in protein synthesis.

## Centrioles

Centrioles are found in all animal cells and most protoctists but are absent from the cells of higher plants. Centrioles arise in a distinct region of the cytoplasm known as the **centrosome** and consist of two hollow cylinders. At cell division they migrate to opposite poles of the cell where they synthesize the microtubules of the spindle.

## Microtubules

These are slender, unbranched tubes made of protein. They have several functions, e.g. they can assemble to form the spindle in cell division.

### Vacuoles

Vacuoles are fluid-filled cavities bounded by a single membrane. The mature plant cell contains a large, permanent, central vacuole filled with chemicals such as glucose and ions in water. This is called the **cell sap**. The membrane around the vacuole is called the **tonoplast**. The vacuoles of animal cells are much smaller and less permanent.

**Checkpoint 4**

Suggest the difference in destination of proteins produced by 'free' ribosomes and those produced by rough ER.

**Links**

Details of cell division (mitosis) are on pages 20 and 21.

**Links**

The vacuole plays an important role in maintaining the internal hydrostatic pressure of the cell as described on page 29.

---

**Exam questions**                              answers: page 26

1   The diagram below shows part of an animal cell and is based on a series of electron micrographs.

(a) Name the structures labelled A–D.

(b) Label with a letter E a structure where the protein contents of D are synthesized.

(4 min)

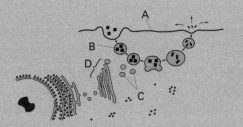

# Chromosome structure and mitosis

You've already learnt that inside the nucleus are chromosomes. Chromosomes contain DNA, which contains hereditary information that is transmitted from cell to cell when cells divide. In any one species the number of chromosomes per body cell nucleus is constant. Mitosis is the division of the nucleus to produce two daughter nuclei containing identical sets of chromosomes.

## Chromosome structure

The number of chromosomes in the cells of different species varies. Humans normally have 46 chromosomes, a mouse has 40 chromosomes. Chromosomes are made up of DNA, protein and a small amount of RNA. DNA occurs as a single strand in the form of a double helix running the length of the chromosome. It is only at the onset of cell division that chromosomes become visible. Each chromosome is seen to consists of two threads called chromatids, which lie parallel along most of their length but are joined only in a specialized region called the **centromere**. The centromere holds the two chromatids together.

## Mitosis

Dividing cells undergo a regular pattern of events, known as the cell cycle. This is a continuous process but for convenience of description it is subdivided into four stages plus a 'resting' stage, known as interphase, between one complete division.

### Interphase
This is the longest part of the cycle during which a newly formed cell increases in size and produces organelles lost during the previous division. Just before the next cell division the chromosomes replicate so that each then consists of two chromatids joined together by the centromere.

### Prophase
This is the phase of mitotic division during which the chromatids shorten and become thicker by the coiling and condensation of the DNA protein coat. In the cells of animals and lower plants, the centrioles move to the poles of the cells and microtubules begin to radiate from them forming asters. At the end of prophase the nuclear membrane disintegrates and the spindle is formed. During this phase the nucleoli disappear.

### Metaphase
The chromosomes are arranged at the centre or equator of the spindle and become attached to certain spindle fibres at the centromere. Contraction of these fibres draws the individual chromatids slightly apart.

**Check the net**

You'll find up-to-date information on mitosis at buglady.clc.uc.edu/biology/bio104/mitosis.htm
www.biology.arizona.edu

**Checkpoint 1**

Name the point at which pairs of chromatids are held together.

**Examiner's secrets**

Be prepared to compare meiosis and mitosis, and don't confuse the two. (Meiosis is described on pages 142 and 143.)

**Speed learning**

Memorize the correct order of the stages of mitosis by making up a mnemonic, e.g. Important People Make Awful Tea.

**Action point**

Construct a table summarizing the changes that occur at each of the four stages of mitosis.

**Checkpoint 2**

Name the stage at which the DNA content doubles.

## Anaphase

This stage is very rapid. The centromere splits and the spindle fibres pull the now separated chromatids to the poles of the cell, where they become the chromosomes of the two daughter cells.

## Telophase

Mitosis ends with telophase. The chromosomes, having reached the poles of the cells, uncoil and lengthen. The spindle breaks down, the centrioles replicate, the nucleoli reappear and the nuclear membrane re-forms. In animal cells cytokinesis occurs by the constriction of the centre of the parent cell from the outside inwards. In plant cells, a cell plate forms across the equator of the parent cell from the centre outwards and a new cell wall is laid down.

## Significance of mitosis

→ Mitosis produces two cells which have the same number of chromosomes as the parent cell and each chromosome is an exact replica of one of the originals, i.e. they have identical DNA.

→ The division allows the production of cells that are genetically identical to the parent and so gives genetic stability.

→ By producing new cells, mitosis leads to growth of an organism and also allows for repair of tissues and the replacement of dead cells.

→ An additional function of mitosis is to provide for asexual reproduction, e.g. in plant cuttings, bacteria.

**Checkpoint 3**

Which cell organelles are concerned with spindle formation?

**Checkpoint 4**

In what situation might it be a disadvantage for the offspring to be identical to the parent?

**Test yourself**

A question on mitosis may provide you with drawings of unnamed stages in the incorrect order. You may be required to name the stages and give the correct order.
Draw diagrams of the stages and glue them onto cards. On separate pieces of paper write out the names of the stages. Then lay out the cards in the correct order and label the stages!

**Exam question**                                    answer: page 26

Complete the labels for the diagrams below which show two phases of mitosis. (5 min)

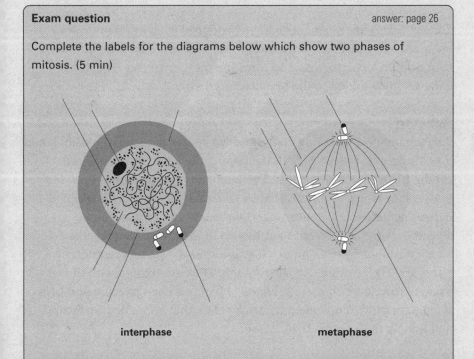

interphase                                    metaphase

# Gene technology

Genetic engineering is the *in vitro* manipulation of DNA. It involves the introduction of engineered DNA into cells in such a way that it will replicate and be passed on to progeny cells. The techniques involved in this work are also being used in the clinical field, with research into changing human DNA (gene therapy), chromosome sequencing and in the forensic field (DNA fingerprinting). One important application is the introduction of DNA from various organisms into bacterial cells, which then produce the required product. This aspect is dealt with in more detail in a later section.

## Recombinant DNA technology ●●●

There are four stages in gene manipulation:

→ formation of DNA fragments including the gene selected for replication
→ splicing (insertion) of the DNA fragments into the vector
→ introduction of the vector into the bacterium
→ selection of the newly transformed organism for cultivation

### Formation of DNA fragments

**Restriction endonucleases** are used to cut DNA between specific base sequences that the enzyme recognizes. For example the *Hin*dIII nuclease recognizes a six base-pair sequence, cutting it as shown below. Most restriction enzymes split the two strands in a staggered sequence, forming 'sticky ends'.

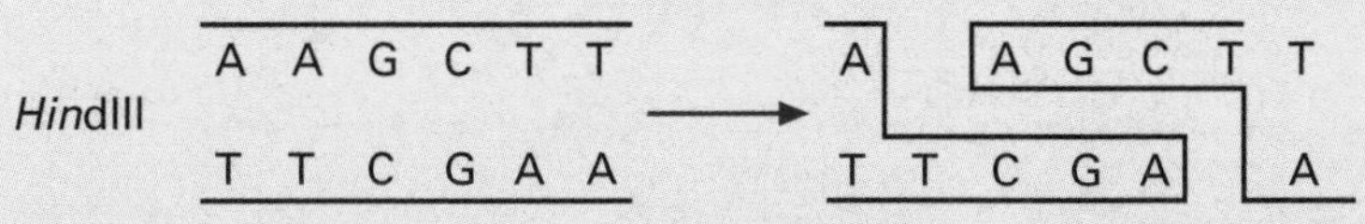

Longer eukaryote fragments can be produced from purified mRNA using the enzyme **reverse transcriptase** to produce a single strand of DNA and DNA polymerase 1 to form the partner strand. The sticky ends are then added to the manufactured strand.

### Splicing

Splicing involves breaking open the DNA ring of a bacterial plasmid, and inserting a short piece of DNA containing the gene for the desired product from a donor species. The manufactured **plasmid** pBR322 is most used, as in the transformation of *Escherichia coli*. This plasmid carries genes for resistance to certain antibiotics and so can be selected by the use of these antibiotics in the growth media. It is a small plasmid and has unique cleavage sites for several restriction enzymes. The plasmid and the required DNA are treated with the same restriction enzyme, e.g. *Hin*dIII, giving an open plasmid and DNA fragments with complementary sticky ends that are joined by mixing and adding the enzyme **DNA ligase**.

**Checkpoint 1**

Why are prokaryotes more appropriate for recombinant DNA technology than eukaryotes?

**Checkpoint 2**

Name the specific biological molecule that forms the plasmid.

**The jargon**

The *sticky ends* are the two ends of the foreign DNA segment, which have a short row of unpaired bases that match the complementary bases at the two ends of the opened-up plasmid.

**Checkpoint 3**

Name the enzyme that splits the DNA molecule.

**The jargon**

A *plasmid* is a small circular loop of DNA found in bacterial cells. Bacterial plasmids are widely used as vectors.

Once a bacterium has taken up a piece of foreign DNA successfully, the foreign DNA replicates along with the rest of the plasmid every time the bacterial cell divides. The bacterium may divide repeatedly and give rise to a large population of bacterial cells all of which contain replicas of the foreign DNA.

## Applications of genetic engineering

Two main applications are considered:

→ health care
 → production of human hormones, e.g. insulin, growth hormones
 → production of antibiotics and vaccines
 → gene therapy, where cells containing non-mutant genes are substituted for the abnormal mutant cells
→ agriculture
 → genetically engineered hormones are used to increase meat yield
 → development of crop plants resistant to herbicides and pests, e.g. by transferring genes that produce toxins with insecticidal properties from bacteria to higher plants such as potatoes

## Implications of genetic engineering

It is impossible to predict what the consequences might be of releasing genetically engineered organisms into the environment. The potential hazards are:

→ a new gene, on insertion, may disrupt normal gene function
→ a potentially dangerous microorganism with a new gene may become a dangerous pathogen if it is released into the environment
→ the recombinant DNA might get into other organisms, e.g. herbicide resistance might be transferred to a weed species

---

**Action point**

Draw a flowchart showing how a plasmid vector may be used in recombinant DNA technology.

**Links**

See pages 152 and 153 for further applications.

**Checkpoint 4**

*E. coli* is a bacterium used to clone genes. It is widely distributed and capable of exchanging genetic material with other types of bacteria. Suggest why some scientists think that the production of recombinant DNA could be dangerous.

**Action point**

Make a list and give definitions for the words in bold on page 22.

---

**Exam question**                                    answer: page 26

Read the following passage on genetic engineering and complete using the most appropriate word(s). 'The isolation of specific genes during a genetic engineering process involves forming eukaryotic DNA fragments. These fragments are formed using .................... enzymes which make staggered cuts in the DNA within specific base sequences. This leaves single-stranded "sticky ends" at each end. The same enzyme is used to open up a circular loop of bacterial DNA which acts as a .................... for the eukaryotic DNA. The complementary sticky ends of the bacterial DNA are joined to the DNA fragment using another enzyme called .................... .

DNA fragments can also be made from .................... templates. Reverse transcriptase is used to produce a single strand of DNA and the enzyme .................... catalyses the formation of a double helix. Finally, new DNA is introduced into host .................... cells. These can then be cloned on an industrial scale and large amounts of protein harvested. An example of a protein currently manufactured using this technique is .................... .'

(7 min)

**Test yourself**

Use a flow diagram to illustrate your answer to the following essay question: 'Explain how genetic engineering can be used to make insulin'.

# Answers
## Molecules and cells

## Molecules 1

### Checkpoints

1 Five.
2 In α glucose the OH group on carbon 1 points downwards whereas the H group points upwards (in β glucose the H group points upwards and the OH group points upwards).
3 Condensation reaction.
4 β glucose.

### Exam questions

1 (a) Plant cell wall.
   (b) (i)   Glucose (no need to specify α or β form).
       (ii)  Hexose.
       (iii) Parallel chains; cross-linked with hydrogen bonds.
       (iv)  Tensile strength (strong) or porous.

## Molecules 2

### Checkpoints

1 The phosphate group is ionized and therefore water soluble at that end of the molecule (hydrophilic).
2 Amino acids dissolve in water but are insoluble in organic solvents.
3 By means of hydrogen bonds.
4 An increase in kinetic energy causes the breaking of hydrogen and ionic bonds.
5 Globular.

### Exam questions

1 (a) (i)  A: amino group.
           B: carboxyl group.
       (ii) H.
   (b) (i)

       (ii)  box on diagram.
       (iii) condensation.

## Nucleic acids

### Checkpoints

1 TCGGATGCA.
2 Organic bases.
3 In DNA the sugar is deoxyribose whereas in RNA the sugar is ribose.
   In DNA the base thymine is present whereas in RNA it is replaced by uracil.
   DNA is double stranded, RNA is single stranded.
4 DNA.

### Exam questions

1 (a) (i)   hydrogen.
       (ii)  adenine or guanine.
       (iii) nitrogen; phosphorus.

(b) Thymine in DNA is replaced by uracil in mRNA.
    DNA is a double helix while mRNA is usually not.
    DNA is double stranded while mRNA is single stranded.
    In DNA the sugar is deoxyribose and in mRNA it is a ribose sugar, or reference to the sugar in RNA containing more oxygen.
    In DNA the bases are paired and in RNA they are not.
    In DNA the ratio of adenine + thymine/cytosine + uracil is about equal while in RNA the ratio of adenine + uracil/cytosine + guanine is more variable.

## Replication and protein synthesis

### Checkpoints

1 672/3 = 224.
2 Ribosome or rough ER.
3 mRNA.
4 tRNA.
5 DNA.
6 tRNA.
7 mRNA.

### Exam question

Your answer should include a detailed description of protein synthesis (see text) and should include diagrams.

### Examiner's secrets

This is a good example of where planning is essential. List the different forms of nucleic acid, DNA and the three forms of RNA. Then list the stages in protein synthesis: transcription, amino acid activation and translation.

## Enzymes 1

### Checkpoints

1 The number of substrate molecules which an enzyme can act upon in a given time is called its 'turnover number'. Catalase has a turnover number of several million, the highest known turnover number.
2 The site on the enzyme molecule where the substrate molecule is held and where the reaction occurs.
3 With increased temperature there is increased kinetic energy of the molecules, i.e. both substrate and enzyme molecules move about more rapidly and therefore the chance of collisions is increased. However, when a temperature of about 45 °C is reached denaturation occurs (hydrogen bonds are broken as they are individually weak) and the shape of the protein molecule is irreversibly altered. At low temperatures, e.g. 0 °C, there is little movement of molecules and enzymes are said to be inactive. However, since no hydrogen bonds are broken the shape remains unaltered and so a rise in temperature causes an increase in the rate of the reaction as the enzyme molecule has not been damaged.

**4** No effect.

**5** With immobilized enzymes:
- the product is not contaminated with enzyme
- enzyme is protected against changes in pH/temperature
- several enzymes with differing pH/temperature optima can be used together (a wider range of conditions can be tolerated)

**6** Enzymes are more efficient, i.e. have a higher turnover number and are extremely specific. However, they are denatured by heat and affected by pH.

## Exam question

Your essay should cover all the following points in a logical sequence.
- Most reactions require energy to induce them to occur.
- This increases the activity of molecules, causing reactants to collide (or enhances break-up).
- The combination of reactants with enzymes reduces the required activation energy.
- This allows reactions to occur at physiological (body) temperatures.
- Enzymes are proteins.
- They are held in tertiary structure (or globular) by intermolecular bonds (hydrogen, disulphide, etc.) or the shape is held by bonding.
- This produces a specifically shaped active site.
- This allows the formation of an enzyme–substrate complex.
- You should provide a diagram of the lock and key mechanism.

## Enzymes 2

### Checkpoints

**1** The rate of formation of product would increase. This is because the chance of an enzyme–substrate molecule contact is now greater than the chance of an enzyme–inhibitor contact.

**2** A competitive inhibitor attaches to the active site whereas a non-competitive inhibitor attaches at a point other than the active site.

**3** Competitive inhibition.

**4** Non-reversible inhibition leaves the enzyme permanently damaged whereas if the reversible inhibitor is removed the enzyme can work normally once more.

### Exam questions

**1** (a) (i)  curve A.
- The concentration of substrate limits the rate at low concentrations **or** at low concentrations there is more substrate for enzymes to work on.
- At high concentrations the enzyme is the limiting factor because the enzyme is working as fast as possible **or** there is not enough enzyme molecules present to combine with all the substrate molecules.

(ii) With the competitive inhibitor (curve B) there is an increased chance of the enzyme meeting the substrate. With non-competitive inhibitor (curve C) the enzyme is permanently inactivated.

(b) (i)  The active site changes shape to accommodate the substrate.

(ii) With the lock and key theory it is not possible to alter the shape of the active site whereas the induced fit theory means the active site can change shape.

**2** (a) Hydrogen/sulphur bonds are broken.
This alters the tertiary structure of enzymes **or** enzymes are denatured.
This changes the shape of the active site so that the substrate no longer fits.

(b) Using figures from the table, above 40 °C the rate of reaction is higher in immobilized enzymes.

(c) Trapping **or** encapsulation **or** bonding **or** adsorption.

## Cell structure and cell membrane

### Checkpoints

**1** Prokaryote: no nuclear membrane, circular strands of DNA, no membrane-bound organelles, no meiosis/mitosis. Eukaryote: membrane-bound nucleus, chloroplasts and mitochondria may be present, meiosis and mitosis occur.

**2** Plasma membrane/cell membrane.

**3** Cell wall.

**4** Extrinsic proteins are found in or on only one of the layers of the membrane whereas intrinsic extend across both layers.

**5** As opposed to static, dynamic suggests movement, i.e. the protein floats about, although some proteins are anchored to organelles within the cell. The lipid component also moves.

**6** Fat-soluble molecules.

**7** In between the phospholipid molecules.

### Exam questions

**1** (a)
- The membrane consists mainly of phospholipid molecules which have a hydrophilic head and a hydrophobic tail.
- Give a description or diagram showing glycerol joined to two fatty acids and one phosphate molecule.
- The glycerol and phosphate is hydrophilic.
- The fatty acid part is hydrophobic.
- Since both sides of the plasma membrane are in contact with aqueous solutions the hydrophilic heads are orientated both inwards and outwards. forming a bilayer with the tails in the centre.

(b)
- Membrane phospholipid molecules have weak intermolecular forces and can move freely relative to one another.
- There are 'floating islands' of protein held in a membrane.

- Some (intrinsic) proteins pass right through the bilayer.
- Some (extrinsic) proteins float in the outer half of the bilayer or are attached to its surface.

## Cell organelles

### Checkpoints

1 To allow mRNA to pass through from the nucleus to the cytoplasm.
2 Chloroplast.
3 Mitochondrion.
4 Free ribosomes produce proteins which are retained in the cell, whereas ribosomes attached to ER produce proteins which are transported or secreted outside the cell.

### Exam questions

1 (a) A: plasma membrane **or** plasmalemma **or** cell membrane.
 B: phagocytic vesicle **or** food vacuole.
 C: lysosomes **or** Golgi vesicle.
 D: Golgi apparatus.
 (b) Label ribosome on rough ER.

## Chromosome structure and mitosis

### Checkpoints

1 Centromere.
2 Interphase.
3 Centrioles.
4 If the environment changes there will be no variants that might be more suited to the environment.

### Exam question

This answer provides labelling only. You should refer to your notes for descriptions of phases.

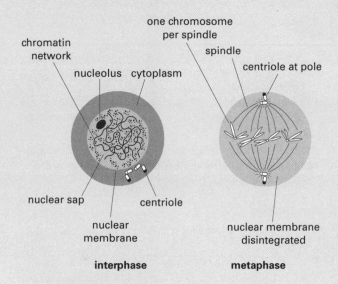

interphase          metaphase

### Examiner's secrets

The examiner has kindly named the phases for you! All you need to do is label the diagram correctly. You *could* be asked to describe the key characteristics of each phase or describe the differences between the phases.

## Gene technology

### Checkpoints

1 All the genes of prokaryotes are expressed in the next generation since the DNA is single stranded.
2 DNA.
3 Restriction endonuclease.
4 Introduced DNA could spread among bacterial, plant and animal populations with unpredictable results.

### Exam question

Restriction endonuclease; vector; DNA ligase; mRNA; DNA polymerase; bacterial; insulin/growth hormone/interferon.

# Exchange and transport

All living organisms exchange gases with the environment. The raw materials for processes like respiration and photosynthesis must be transported to the cells which need them and waste products must be carried away. Soluble food absorbed by the gut wall of animals must be transported around the body, and the products of photosynthesis must be taken to all parts of the plant. Materials also need to pass in and out of cells as they are required and produced by the various cell organelles.

## Exam themes

How membranes are involved in the movement of materials in and out of cells

Methods of movement of molecules

The structure of the human gut and associated organs

The structure of the ileum and its functions in digestion and absorption

The opening and closing mechanism of stomata

The rate of respiration and its nervous control

The control of ventilation

The effect of activity on the respiratory cycle

Functions of blood in the transport of gases

The significance of the dissociation curves of adult oxyhaemoglobin at different carbon dioxide levels

The relationship between the structure and function of arteries, veins and capillaries

The structure of red and white blood cells and the differences between blood, tissue fluid and lymph

The structure and function of the mammalian heart

Heart rate and the control of heartbeat

The structure of xylem and phloem in relation to their transport functions

The pathway of water transport

The mass flow theory of translocation in the phloem

Evidence for transport in phloem

Factors affecting the rate of transpiration

Adaptations of plants to reduce water loss

## Topic checklist

| ○ AS  ● A2 | AQA/A | AQA/B | EDEXCEL | OCR | WJEC |
|---|---|---|---|---|---|
| Movement of molecules | ● | ○ | ○ | ○ | ○ |
| Human digestion | ● | ○ | ○ | ● | ○ |
| Gas exchange in plants and simple organisms | ● | ○ | ○ | ○ | ○ |
| Gas exchange in mammals | ○● | ○ | ○ | ○ | ○ |
| Transport in mammals: blood | ○● | ○ | ○ | ○ | ○ |
| Transport in mammals: vessels | ○● | ○ | ○ | ○ | ○ |
| Transport in mammals: heart | ○● | ○ | ○ | ○ | ○ |
| Transport in plants | ● | ○ | ○ | ○ | ○ |
| Transport of water through the plant | ● | ○ | ○ | ○ | ○ |
| Translocation in plants | | ○ | ○ | ○ | ○ |

# Movement of molecules

The organelles and structures within a cell require a variety of materials in order to carry out their functions. In turn they form products, some useful and some waste. To understand plants and animals as whole organisms you first need to look at how materials enter and leave cells.

**Checkpoint 1**

Suggest how an increase in temperature would affect the rate of diffusion.

## Diffusion

Diffusion is the movement of molecules or ions from a region where they are in high concentration to a region of lower concentration. Ions and molecules are always in a state of random movement, but if they are highly concentrated in one area there will be a net movement away from that area until equilibrium is reached or until there is a uniform distribution. The rate of diffusion depends on:

→ the concentration gradient – a steeper gradient means a faster rate
→ the distance over which it takes place – the shorter the distance the greater the rate
→ the surface area – the larger the surface area the greater the rate
→ temperature – important in plants and cold-blooded animals

**Links**

Remember the structure of the cell membrane given on pages 16 and 17.

## Facilitated diffusion

Charged particles or ions and large molecules such as glucose do not readily pass through the cell membrane because they are relatively insoluble in lipid. In the cell membrane protein molecules span the membrane from one side to the other and help such particles to diffuse in or out of the cells. These proteins are of two types.

**Checkpoint 2**

Explain how the structure of the cell membrane determines the preferential uptake of lipid-soluble molecules compared to water-soluble molecules.

→ *Channel proteins*: these form water-filled pores in the membrane. As the channel is hydrophilic, water-soluble ions can pass through. The channels are selective and in this way the cell can control the entry and exit of molecules and ions.
→ *Carrier proteins*: these bind molecules to them and then change shape as a result of this binding in such a way that the molecules are released on the other side of the membrane. They conduct molecules in either direction and energy from ATP is not needed.

**Checkpoint 3**

What would happen to the appearance of cells placed in concentrated glucose solution?

**Watch out!**

Water potential is always expressed as a negative value.

## Osmosis

In biological systems osmosis is a special form of diffusion which involves the movement of water molecules. Most cell membranes are permeable to water and certain solutes only. These membranes are termed partially permeable. The term 'water potential' (WP) is used by biologists to describe the tendency of water molecules to move from high concentrations of water molecules to lower concentrations, i.e. water will move from a region of high WP to one of lower WP. This is because where there is a high concentration of water molecules they have a greater potential energy. A higher WP implies a greater tendency to leave. The WP of pure water is zero, so all solutions have lower WPs than pure water and are given negative values. Water will diffuse from a region of less negative (higher) WP to one of more negative (lower) WP.

**Checkpoint 4**

Cell A has a WP of −10 kPa and cell B has a WP of −20 kPa. In which direction will water move?

If a plant cell is surrounded by pure water the water will flow by osmosis across the cell membrane and cytoplasm and into the vacuole. The vacuole swells causing a pressure against the cell wall. The internal hydrostatic pressure increases until the cell can take in no more water and the cell is said to be turgid.

## Active transport ●●●

Unlike the processes described so far, active transport is an energy-requiring process in which ions and molecules are moved across membranes against a concentration gradient.

→ Ions and molecules can move in the opposite direction to that occurring in diffusion, i.e. against a concentration gradient.
→ The energy is supplied by ATP, and anything which affects the respiratory process will affect active transport.
→ The process occurs through the carrier proteins that span the membrane.
→ The proteins accept the molecule and then the molecule enters the cell by a change in shape of the carrier molecule.

**Exam questions**                                    answers: page 48

1  (a) Active transport, facilitated diffusion and osmosis are three of the ways in which substances move into cells. Complete the table by ticking the statements that are true for the particular method of transport.

| Statement | Active transport | Facilitated diffusion | Osmosis |
|---|---|---|---|
| Requires energy from ATP |  |  |  |
| Involves protein carrier molecules |  |  |  |
| Involves movement of molecules from where they are in a high concentration to where they are in a lower concentration |  |  |  |

*Method of transport into cell*

(b) A turgid plant cell was placed in a solution of sucrose. The diagram below shows the appearance of the cell after 1 hour.

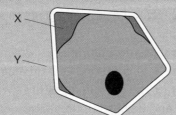

X

Y

(i) From the diagram, what is the evidence which shows that the water potential of the cell sap must be higher than that of the sucrose solution?

(ii) Explain why the water potential at point X is equal to that at point Y.

(6 min)

**Checkpoint 5**

What is meant by the terms 'hypertonic', 'hypotonic' and 'isotonic'?

**The jargon**

Turgor (being turgid) plays an important role particularly in young and herbaceous plants. It supports them and maintains their shape and form.

**Action point**

Draw annotated diagrams to show the effect of immersing a partially turgid plant cell in
(a) pure water
(b) a solution with a solute concentration exceeding that of the cell sap in the vacuole.

**Links**

See mineral salt uptake by roots on page 45.

**Checkpoint 6**

How will an increase in temperature affect active transport?

**Don't forget**

Any practical work connected with water potential and plasmolysis of cells.

# Human digestion

A heterotrophic organism cannot synthesize its major food requirements and so depends on a source of complex organic molecules – in other words, it needs to eat! The food is required as a source of energy for activities such as movement, and for the synthesis of body tissues. The organic molecules must be broken down by digestion and absorbed into the body tissues from the digestive system before being used in the body cells. You need to study what happens in each part of the alimentary canal.

## Digestion ●●●

The absorption of nutrients by the gut epithelial cells is only possible if the large molecules, carbohydrates, fats and proteins, are first broken down or digested into smaller products by means of enzymes. At the completion of digestion carbohydrates are broken down to monosaccharides, proteins to amino acids, and fats to fatty acids and glycerol.

### Mouth

Digestion starts in the mouth, with chewing of the food by the teeth. While chewing is in progress, the food is also mixed with saliva from the salivary glands. Saliva is a watery secretion containing mucus and **salivary amylase**, together with some mineral ions. It is important for lubricating the food before it is swallowed. Amylase is an enzyme that breaks down starch to maltose. The food is formed into a bolus, which passes into the oesophagus where it is moved down to the stomach by peristaltic contractions of the longitudinal and circular muscles.

### Stomach

The presence of food in the mouth stimulates the secretion of the **gastric** juice from the gastric glands in the stomach wall. The stomach is a muscular sac that contracts rhythmically and mixes up the food with the gastric juice. The gastric glands are simple tubular glands with peptic cells, oxyntic cells and mucus-secreting cells.

→ *Peptic cells* secrete the protein-digesting enzyme pepsin as an inactive precursor, pepsinogen. This prevents the enzyme from damaging the stomach tissues before it is released. When it is activated, it is prevented from damaging the stomach wall by the secretion of mucus.

→ *Oxyntic cells* secrete hydrochloric acid, which makes the stomach contents acid; the acid conditions kill off many pathogenic bacteria as well as activating the protein-digesting enzymes; pepsinogen becomes pepsin and begins the breakdown of proteins to polypeptides; pro-rennin becomes rennin, which coagulates the soluble protein in milk, so this enzyme is especially important in young mammals.

**Examiner's secrets**

You do not have to learn about all the enzymes involved in digestion.

**Checkpoint 1**

Waves of muscular contraction (peristalsis) move across the stomach wall after food arrives there. What part does peristalsis play in digestion?

**Checkpoint 2**

What is the difference between an exocrine and an endocrine gland?

**Checkpoint 3**

Would you classify the pancreas as an endocrine gland, exocrine or both?

**Checkpoint 4**

Why is it essential that digestive enzymes such as proteases be secreted in an inactive state?

**The jargon**

A *precursor* is an inactive enzyme. The precursor is activated only when food is present.

**Links**

Remember the structure of carbohydrates, fats and proteins on pages 4–7.

## Small intestine

The small intestine in humans is divided into two main regions, the duodenum and the ileum. The **duodenum** comprises the first 20 cm of the small intestine and receives secretions from both the liver and the pancreas.

→ Bile is produced in the liver and stored in the gall bladder, from where it passes into the duodenum via the bile duct. It contains no enzymes but the bile salts are important in emulsifying the lipids present in the food. Emulsification is achieved by lowering the surface tension of the lipids, causing large globules to break up into tiny droplets. This enables the action of the enzyme lipase to be more efficient as the lipid droplets now have a much larger surface area. Bile also helps to neutralize the acidity of the food as it comes from the stomach.

→ **Pancreatic** juice is secreted from the exocrine glands in the pancreas and enters the duodenum through the pancreatic duct. It contains a number of different enzymes:

→ *pancreatic amylase*, which breaks down any remaining starch to maltose;

→ *pancreatic lipase*, which splits lipids into fatty acids and glycerol;

→ *trypsin* (secreted as the inactive precursor trypsinogen), which continues the breakdown of proteins.

In the walls of the duodenum are Brunner's glands, which secrete an alkaline juice and mucus. The alkaline juice helps to keep the contents of the small intestine at the correct pH for enzyme action, and the mucus is for lubrication and protection.

Enzymes are secreted from groups of cells at the bottom of the crypts of Lieberkuhn. Here maltase splits maltose into two glucose molecules and endopeptidases and exopeptidases complete the digestion of proteins to amino acids.

## Absorption ●●●

Absorption follows digestion and takes place mainly in the small intestine. The surface available for absorption is greatly increased by the presence of villi.

→ Glucose and amino acids are absorbed across the epithelium of the villi by a combination of diffusion and active transport, and pass into the capillary network that supplies each villus.

→ Fatty acids and glycerol pass into the epithelial cells and recombine to form neutral fat which is then passed into the lymphatic system.

**Checkpoint 5**

What changes in pH occur in the various regions of the gut during the digestion of food?

**Checkpoint 6**

Name two essential constituents of food that are not digested.

**Checkpoint 7**

Name three products of the digestion of bread.

**Test yourself**

For the process of digestion complete a table with the following headings: organ of secretion, site of action, contents, effect.

**The jargon**

*Peptidases* are enzymes that digest proteins and there are two groups. *Endopeptidases* hydrolyze the peptide bonds between amino acids in the central region of the polypeptide. *Exopeptidases* hydrolyze the peptide bonds on the terminal amino acids, progressively reducing them to their individual amino acids.

**Action point**

Draw a detailed section through a villus to show regions of absorption.

**Exam question**                                                         answer: page 48

Describe and illustrate the structure and histology of the mammalian small intestine. Explain fully the ways in which the small intestine is adapted to carry out its functions. (20 min)

# Gas exchange in plants and simple organisms

Gas exchange involves the exchange of oxygen and carbon dioxide between the organism and its environment. It takes place at a respiratory surface by the process of diffusion. You should understand the role of stomata in the exchange of gases in plants.

## The respiratory surface

In order to achieve the maximum rate of diffusion a respiratory surface must have the following features:

→ thin so that diffusion paths are short
→ permeable to the respiratory gases
→ moist to allow easier diffusion of gases
→ a sufficiently large surface area to satisfy the needs of the organism

**Checkpoint 1**

Why must all microscopic organisms live in water?

## Single-celled organisms

In simple, single-celled organisms diffusion of gases occurs over the whole of the body surface. The surface area to volume ratio is large enough to satisfy the organism's needs, the diffusion paths are short and there is no need for a circulatory system.

**Checkpoint 2**

Suggest how organisms such as flatworms overcome their problem of being multicellular without needing a circulatory system.

## Plants

The structure of the leaves of flowering plants (see diagram below) is related to their function of gaseous exchange. To enable gaseous exchange to take place efficiently:

→ the lamina is thin so diffusion paths for gases are short
→ the spongy mesophyll tissue allows for the circulation of gases
→ the plant tissues are permeated by air spaces
→ the stomatal pores permit gas exchange
→ leaves have a large surface area to volume ratio

**Action point**

Draw a labelled diagram of a transverse section of a leaf.

**Checkpoint 3**

Explain why in the daylight plants take in carbon dioxide and give out oxygen and so appear not to respire.

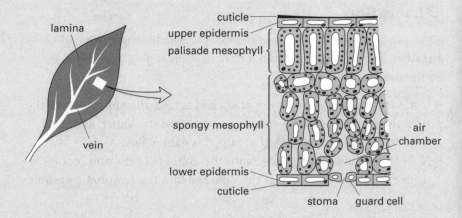

**Action point**

Rephrase the bullet lists as an essay-style paragraph to describe how the leaf is adapted to its function of gaseous exchange.

### Stomata

These are pores in the epidermis, each bordered by two guard cells. The guard cells differ from the other epidermal cells in that they possess chloroplasts. The inner wall of each guard cell is thicker than the outer wall.

Gases diffuse through the stomata along the concentration gradient. Once inside the leaf the gases in the sub-stomatal air chambers diffuse through the intercellular spaces between the mesophyll cells and diffuse into the cells. The direction of diffusion depends on the environmental conditions and the requirements of the plants. It is the net exchange of carbon dioxide and oxygen in relation to respiration and photosynthesis which matters.

Because the inside of the leaf needs to be open to the atmosphere for the exchange of gases it is inevitable that water is lost from the leaves by the process of transpiration. In conditions of water deficit the stomata close.

## The opening and closing mechanism of stomata

During the day the mechanism for stomatal opening occurs as follows.

→ A potassium ion ($K^+$) pump in the cell membranes actively transports $K^+$ ions into the guard cells.
→ As photosynthesis occurs the carbon dioxide concentration falls.
→ The pH rises and an enzyme catalyzes the conversion of starch to malate.
→ $K^+$ and malate ions accumulate in guard cells.
→ The water potential is lowered and water flows in by osmosis.
→ Guard cells become turgid and curve apart more because their outer walls are thinner than the inner walls, so the pore widens (see diagram below).

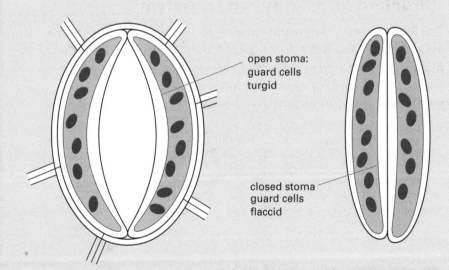

open stoma:
guard cells
turgid

closed stoma
guard cells
flaccid

**Checkpoint 4**

Apart from xylem and phloem cells, which cells of the leaf do not contain chloroplasts?

**Checkpoint 5**

Which three environmental conditions affect the rate of diffusion?

**Links**

Transpiration is discussed on page 43.

**Checkpoint 6**

Which organelle must be in abundant supply for the $K^+$ ion pump to operate?

**Links**

Osmosis is discussed on pages 28 and 29.

**Test yourself**

Explain why stomata close at night.

**Action point**

Describe the mechanism for stomatal **closure**. It is the reverse of the opening mechanism described.

**Exam question**                                    answer: page 48

Outline the functions of the stomata of a leaf. Give an illustrated account of the structure and function of the stomata, describing the theories which have been put forward to explain their opening and closing. (15 min)

# Gas exchange in mammals

Mammals, being relatively large, have a small surface area to volume ratio and often have a higher rate of metabolism than unicellular organisms. They therefore require more oxygen and produce more carbon dioxide. Gases cannot readily penetrate their surface so they, like other multicellular animals, have evolved special gaseous exchange mechanisms.

## Overcoming problems associated with increase in size  ●●●

An increase in size and complexity brings about certain problems:

→ the surface is some distance from most of the cells
→ there is an increase in metabolic rate
→ there is a decrease in surface area to volume ratio

Terrestrial vertebrates have evolved four main ways to overcome these.

→ *Lungs*: a compact respiratory surface but with a large surface area.
→ *A ventilation mechanism*: moves a fresh supply of air over the respiratory surface and maintains diffusion gradients.
→ *An internal transport system*: a blood circulation system moves gases between respiring cells and the respiratory surface.
→ *A respiratory pigment* in the blood increases its oxygen-carrying capacity.

## Structure of the respiratory system  ●●●

The lungs are enclosed within an airtight compartment, the **thorax**, at the base of which is a dome-shaped sheet of muscle called the **diaphragm**. Air is drawn into the lungs via the trachea. The lungs consist of a branching network of tubes called bronchioles arising from a pair of bronchi (see diagram below).

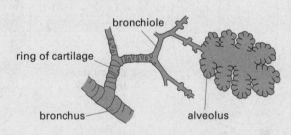

ring of cartilage — bronchiole

bronchus — alveolus

## Ventilation mechanism  ●●●

Breathing in is called inspiration and breathing out is called expiration. Inspiration is brought about as follows. The muscles between the ribs, called the external **intercostal** muscles, contract. At the same time the diaphragm contracts, and flattens. The ribcage moves up and out. This increases the volume of the thorax. The pressure in the thorax falls and air rushes in from the exterior.

**Checkpoint 1**

Name the respiratory pigment found in blood.

**Checkpoint 2**

How is the trachea protected against closure?

**Watch out!**

Be prepared to compare drawings of normal and diseased lung tissue.

**Test yourself**

Describe the process of expiration.

## Gaseous exchange in the alveolus

In the alveoli there is a thin film of moisture in which the oxygen dissolves and then diffuses into the blood capillaries. Carbon dioxide diffuses from the blood into the alveoli (see diagram below).

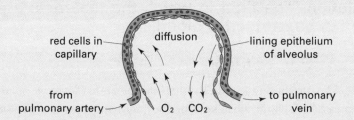

## Rate of respiration

Human lungs have a volume of about 5 dm$^3$, but at rest only about 0.45 dm$^3$ of this will be exchanged. This is called the tidal volume. With increasing activity both the frequency and depth of inspiration will increase. Vital capacity is the total volume of air that can be expired after a maximum inspiration. Even after maximum expiration about 1.5 dm$^3$ of air remains in the lungs. This is called the residual volume.

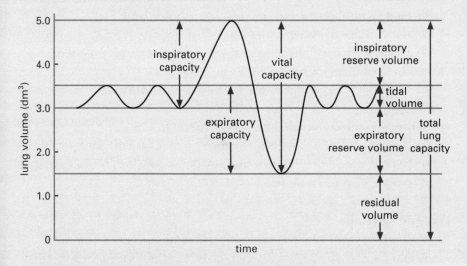

## Nervous control of breathing

Breathing is controlled by a respiratory centre in a region of the hindbrain called the medulla oblongata. The respiratory centre itself is sensitive to pH. Also, chemoreceptors monitor blood pH and therefore detect changing carbon dioxide concentrations. The chemoreceptors are located in:

→ the carotid bodies in the carotid arteries of the neck
→ the aortic body in the wall of the aorta close to the heart

**Links**

The carriage of oxygen and carbon dioxide in the blood is dealt with on pages 36 and 37.

**Checkpoint 3**

List the basic requirements of a respiratory surface.

**Checkpoint 4**

Calculate your ventilation rate (breathing in and out counts as 1).

**Checkpoint 5**

State two effects on your breathing when taking exercise.

**Checkpoint 6**

What does a spirometer measure?

**Action point**

Study the graph opposite and state the volumes of gas exchanged for each of the terms.

**Checkpoint 7**

What is meant by the term 'residual volume'?

---

**Exam question**                                   answer: page 49

Give an illustrated account of the structure of the mammalian respiratory system. (15 min)

# Transport in mammals: blood

In mammals a transport system provides the link between specialized areas for gas exchange and the cells which require oxygen and nutrients. As all cells are bathed in an aqueous medium, the delivery of materials to and from these cells is carried out largely in solution. The fluid in which the materials are dissolved or suspended is blood.

## Composition of blood

Blood is a tissue made up of cells in a fluid plasma. *Plasma* is made up largely of 90% water, with soluble food molecules, waste products, hormones, plasma proteins, mineral ions and vitamins dissolved in it. The cellular elements (blood cells and cell fragments) are of three types:

→ **erythrocytes** or red blood corpuscles
→ **leucocytes** or white blood corpuscles
→ **thrombocytes** or platelets, which are fragments of cells

## Functions of blood

→ Plasma transports digested food products, hormones, proteins, fibrinogen, antibodies, etc. and also distributes heat.
→ Leucocytes are of two groups:
  → granulocytes are phagocytic, have granular cytoplasm, lobed nuclei and engulf bacteria;
  → agranulocytes produce antibodies and antitoxins, have clear cytoplasm and spherical nuclei.
→ Thrombocytes are involved in blood clotting.
→ Erythrocytes are filled with the pigment haemoglobin, are biconcave in shape and do not contain a nucleus. Their function is the carriage of oxygen, described below.

### Oxygen carriage

An efficient respiratory pigment readily picks up oxygen at the respiratory surface and releases it on arrival at the tissues. Respiratory pigments have a high affinity for oxygen when the concentration is high but a low affinity when the concentration is low. When **haemoglobin** is exposed to a gradual increase in oxygen partial pressure it absorbs oxygen rapidly at first but more slowly as the partial pressure increases. This relationship is known as the oxygen dissociation curve (see graph on next page).

The release of oxygen from haemoglobin is facilitated by the presence of carbon dioxide, a phenomenon known as the **Bohr effect**.

→ When the partial pressure of oxygen is high, as in the lung capillaries, oxygen combines with haemoglobin to form oxyhaemoglobin.
→ When the partial pressure of oxygen is low, as found in the respiring tissues, then oxygen dissociates from haemoglobin.
→ When the partial pressure of carbon dioxide is high, haemoglobin is less efficient at taking up oxygen and more efficient at releasing it.

**Checkpoint 1**

Red blood cells live for only three months. Where are they recycled?

**Checkpoint 2**

Why are red blood cells biconcave in shape?

**Test yourself**

Make up a table of the functions of the parts of the blood.

**Test yourself**

Draw labelled diagrams of red and white blood cells.

**The jargon**

Oxygen is measured by *partial pressure*, otherwise called *oxygen tension*. As oxygen makes up around 21% of the atmosphere the partial pressure of oxygen in the atmosphere is around 21 kPa.

**Examiner's secrets**

Answers to questions on the 'Bohr effect' are concerned with: the *uptake* of oxygen by haemoglobin even at low partial pressures and the *offloading* of oxygen by the presence of high levels of carbon dioxide at the tissues.

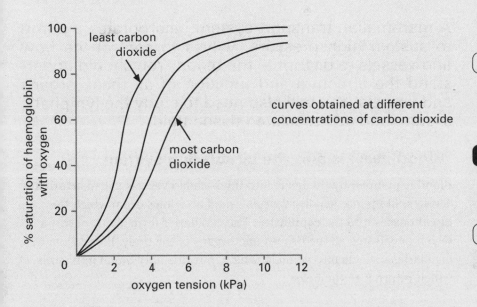

y-axis: % saturation of haemoglobin with oxygen
x-axis: oxygen tension (kPa)

least carbon dioxide

most carbon dioxide

curves obtained at different concentrations of carbon dioxide

## Carbon dioxide carriage

Carbon dioxide is transported in blood cells and plasma in three ways: 85% as hydrogencarbonate, 5% in solution in the plasma and 10% in combination with haemoglobin to form carbaminohaemoglobin.

→ Carbon dioxide diffuses into the red blood cell (RBC) and combines with water to form **carbonic acid**.
→ Carbonic acid dissociates into $H^+$ and $HCO_3^-$ ions, the reaction being catalysed by **carbonic anhydrase**.
→ $HCO_3^-$ ions diffuse out of the RBC into the plasma where they combine with $Na^+$ ions from the dissociation of sodium chloride to form sodium hydrogen carbonate.
→ $H^+$ ions encourage the **oxyhaemoglobin** to dissociate into oxygen and haemoglobin.
→ $H^+$ ions are buffered by their combination with haemoglobin and the formation of haemoglobinic acid (HHb).
→ The oxygen diffuses out of the RBC into the tissues.
→ To balance the outward movement of negatively charged ions, chloride ions diffuse in. This is known as the chloride shift, which maintains the electrochemical neutrality of the RBC.

answers: page 49

**Exam questions**

1  The diagram below shows exchange taking place between a RBC and the fluid surrounding active muscle tissue.
   (a) Name X.
   (b) Name an organ in the body where this pattern of exchange would not occur.
   (c) Explain fully how the release of oxygen is brought about.
   (d) Suggest a reason for the movement of the chloride ions shown in the diagram.

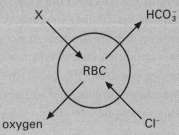

X          $HCO_3^-$

RBC

oxygen          $Cl^-$

(9 min)

---

**The jargon**

The *Bohr effect* explains the way that oxygen moves from the red blood cell to the tissues. It is represented by a graph.

**Checkpoint 3**

What is the effect of the dissociation curve being displaced to the left?

**Watch out!**

Be prepared to explain the significance of the difference between the haemoglobin association curve of a fetus with that of an adult.

**Checkpoint 4**

How does the haemoglobin of an animal with a high metabolic rate differ from that of human haemoglobin?

**Checkpoint 5**

What is the main purpose of the inward movement of chloride ions?

**Test yourself**

Draw a large diagram of a red blood cell and devise a flow chart showing the process known as the chloride shift.

# Transport in mammals: vessels

A mammalian transport system incorporates a pump to sustain high pressure, valves to control the flow and vessels to distribute the blood. You should understand the structure and function of all these organs and vessels. You will also need to study the lymphatic system and its associated tissue fluid.

## Blood vessels and the circulation system ●●●

Blood is pumped by the heart into thick-walled vessels called **arteries**. These split up into smaller vessels called arterioles, from which the blood passes into the **capillaries**. The capillaries form a vast network which penetrates all the tissues and organs of the body. Blood from the capillaries collects into venules, which in turn empty blood into **veins**, which return it to the heart.

### Structure of arteries and veins

The arteries and the veins have the same basic three-layered structure (see diagram opposite) but the proportions of the different layers vary. In both:

→ the *innermost layer*, the endothelium, is one cell thick and provides a smooth lining with minimum resistance to the flow of blood
→ the *middle layer*, which is made up of elastic fibres and smooth muscle, is thicker in the arteries than in the veins
→ the *outer layer* is made up of collagen fibres that are resistant to overstretching

Veins have valves in them but these are not present in arteries apart from the aortic valves. The capillaries are thin-walled, consisting only of a layer of endothelium so that they are permeable to water and dissolved substances such as glucose. It is in the capillaries that the exchange of materials between the blood and the tissues takes place.

## Intercellular fluid and lymph ●●●

When blood reaches the arterial end of a capillary it is under pressure because of the pumping action of the heart and the resistance to blood flow of the capillaries. The hydrostatic pressure forces the fluid part of the blood through the capillary walls into the spaces between the cells. This fluid is called tissue fluid. At the arterial end of the capillary bed it supplies the cells with oxygen and nutrients. In addition at the venous end, tissue fluid picks up carbon dioxide and other excretory substances. Blood pressure, diffusion and osmosis are the forces responsible for the movement of water and solutes into and out of the capillaries (see diagram at top of opposite page). Some of this fluid passes back into the capillaries, but some drains into the **lymphatic system** and is returned to the blood eventually via the thoracic duct, which empties into the subclavian vein.

---

**Checkpoint 1**

Why do arteries need a thick muscular wall?

**Action point**

Draw labelled diagrams of an artery, vein and capillary.

**Action point**

Make up a table of similarities and differences between arteries, veins and capillaries.

**Checkpoint 2**

A capillary network or bed slows down the flow of blood at the tissues. Why is this important?

**Checkpoint 3**

How does blood flow back to the heart?

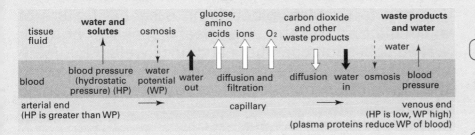

Lymph is the tissue fluid that drains into blind-ending lymphatic capillaries among the tissues that join up to form larger vessels. It is moved by contractions of the muscles through which the vessels pass, being prevented from back-flow by numerous valves similar to those found in the veins. There are lymph glands and nodes associated with the lymph vessels, and these play an important role in the formation of lymphocytes and the prevention of infection.

**Action point**

Explain the forces involved in the movement of water and solutes into and out of the capillaries.

**Links**

See osmosis on pages 28 and 29.

**Checkpoint 4**

What are the two functions of lymph?

**Checkpoint 5**

What components of the blood are normally not found in tissue fluid?

**Watch out!**

Pressure in blood vessels is highest in the arteries, the capillaries are the next highest in pressure and the veins have the least pressure in them. Can you explain why there are these differences in pressure?

**Exam questions**                                   answers: page 49

1   The graph below shows the relationship between blood pressure, blood velocity, and total cross-sectional area for the different types of blood vessels in the mammalian circulatory system.

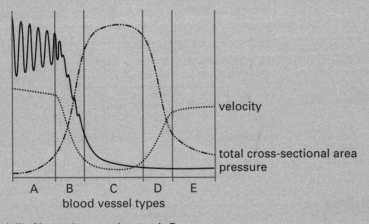

(a) (i)  Name the vessel types A–E.
    (ii) Use the appropriate letter to indicate:
        1   A type of vessel which contains the highest proportion of muscle fibre in its wall.
        2   The vessels which contain most of the total blood volume.

(b) (i)  Account for the variation in blood pressure observed in vessel type A.
    (ii) Suggest a reason for the rapid fall in blood pressure in vessel type B.

(c) Comment on the physiological advantages of the relationship between blood velocity and cross-sectional area of vessel type C.

(d) Suggest two reasons for the increase in blood velocity in vessel types D and E when the blood pressure is so low.

                                                    (12 min)

2   (a) Outline the main chemical components of mammalian blood plasma, briefly indicating one function of each.

    (b) Explain briefly how the blood capillaries exchange other materials with the cells of the body's tissues.

                                                    (15 min)

# Transport in mammals:
## heart

A pump to circulate blood is an essential feature of a circulatory system. The heart consists of thin-walled collection chambers and thick-walled pumping chambers which are partitioned into two, allowing the complete separation of oxygenated and deoxygenated blood.

## The heart ●●●

The heart is situated in the thorax between the two lungs.

→ It is a four-chambered pump as shown in the diagram below.
→ It consists largely of **cardiac** muscle, a specialized tissue that is capable of rhythmical contraction and relaxation over a long period without fatigue.

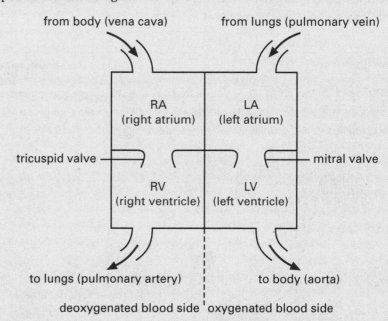

### Blood flow

→ Deoxygenated blood enters the right side of the heart from the **venae cavae** and passes into the thin-walled right atrium.
→ The right atrium contracts and forces the blood into the right ventricle through the **tricuspid** valves.
→ The ventricle contracts forcing blood into the **pulmonary** arteries.
→ The same happens on the left-hand side since both sides of the heart contract simultaneously but the left ventricle wall is much thicker and a greater force is generated when it contracts, forcing the blood into the **aorta** and then into general circulation.

## Control of heartbeat ●●●

The events known as the cardiac cycle can be summarized as follows.

→ The heart muscle is **myogenic**, i.e. the heartbeat is initiated from within a specialized part of the muscle itself and is not due to nervous stimulation.

→ In the wall of the right atrium is a region of specialized cardiac fibres called the sinoatrial node (SAN) which acts as a pacemaker.

→ A wave of electrical stimulation arises at this point and then spreads over the two atria causing them to contract more or less at the same time.

→ The stimulation reaches another specialized region of cardiac fibres, the atrioventricular node (AVN), which lies between the two atria and passes on the excitation to specialized tissues in the ventricles.

→ The excitation passes along the bundle of His and then spreads through the Purkinje tissue in the walls of the ventricles.

→ As with the atria, stimulation is followed by contraction and the walls of the ventricles contract simultaneously.

## Heart rate

The rate at which the heart beats and the volume of blood pumped at each beat can be varied.

cardiac output = volume pumped × number of beats in a given time

Changes to the cardiac output are effected through the autonomic nervous system:

→ impulses from the parasympathetic fibres slow down the heartbeat

→ impulses from the sympathetic fibres accelerate it

The cardiac centre of the brain originates the sympathetic and parasympathetic impulses.

**Action point**

Draw an annotated diagram of the heart showing the control of the heartbeat.

**The jargon**

The *bundle of His* is continuous with the AVN and is a strand of modified cardiac muscle fibre that fans out over the walls of the ventricles.
The *Purkinje* tissue is a network of fibres just beneath the endothelial lining.

**Checkpoint 5**

How many times a day on average does the heart beat?

---

**Exam questions**                                   answers: page 49

1   The diagram below shows the relationship between the heart and part of the nervous system that controls heartbeat.

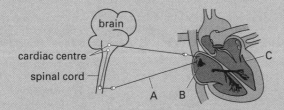

(a) Name and state the function of each of the structures labelled A, B and C.

(b) The cardiac centre of the brain plays an important part in the control of heartbeat.

　(i)   Name the region of the brain where the cardiac centre is located.

　(ii)  Chemoreceptors in the carotid artery respond to increases in carbon dioxide concentration in the blood by stimulating the cardiac centre. Describe the changes brought about by this response.

(c) Patients in which the heart's pumping mechanism has become weak as a result of heart failure are often prescribed diuretics. These remove excess body water. Suggest one reason why this treatment makes it easier for the heart to function.

(15 min)

**Test yourself**

Mammals have a double circulation system, i.e. the blood moves through the heart twice during each cardiac cycle. Draw a section through the heart and from memory label the chambers and associated blood vessels. Use a coloured pen to show the circulation of blood through the heart.

# Transport in plants

In plants the organs collecting the water, the roots, are some distance from the leaves, which require the water for the process of photosynthesis. A transport system through the stem is necessary to connect the two organs. You should study the distribution of xylem and phloem tissues in stems, roots and leaves, and the process by which water is lost through the leaves.

## Vascular tissues

The vascular tissues are made up of:

→ **xylem**, which transports water and mineral salts from the roots to the leaves
→ **phloem**, which mainly transports the soluble products of photosynthesis from the leaves to the other parts of the plant

## Xylem distribution

The distribution of xylem tissue differs in primary stems, leaves and roots.

→ In *stems* it occurs as part of the peripheral vascular bundles. This gives flexible support, but also resistance to bending strain.
→ In *leaves* the arrangement of vascular tissues in the midrib and the network of veins gives flexible strength and resistance to tearing strains.
→ In *roots* the central arrangement is ideal for resistance to pull and so helps the anchorage of the plant.

The different arrangements of xylem tissue are shown in the diagrams below.

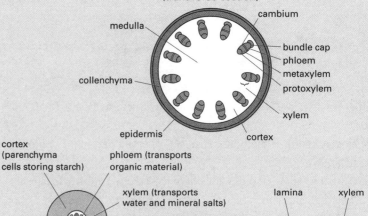

Low power map of tissues of young sunflower stem (transverse section)

TS dicotyledon root (*Ranunculus*) primary structure

TS dicotyledon leaf (a typical mesophytic leaf)

**Watch out!**

Don't confuse *xylem* and *phloem*.

**Checkpoint 1**

State two differences between the arrangement of tissues in a stem and a root.

**Checkpoint 2**

Describe how the arrangement of vascular tissue in the stem, root and leaf enables each to overcome the effect of the wind as a mechanical factor.

**Test yourself**

From memory draw labelled diagrams of TS stem and root.

# Transmport of water
## Transpiration ●●●

The rate at which water is lost from plants is called the transpiration rate and is dependent on external factors such as temperature, humidity and air movements. A number of anatomical and morphological features of plants also influence the transpiration rate. These include the numbers and distribution of stomata, leaf area and thickness of the cuticle.

Some plants exhibit **xeromorphic** adaptations and are known as xerophytes. They need to prevent excessive water loss for one of a number of reasons. They may live in:

→ hot, dry desert regions
→ cold regions where the soil water is frozen for much of the year
→ exposed, windy situations

These xeromorphic adaptations may take a number of general forms:

→ *reducing transpiration rate* by having
  → rolled leaves
  → thick cuticle
  → depression of stomata
  → absence of leaves
  → reduction in surface area/volume ratio of leaves
→ *storing water* (succulents)
  → succulent leaves and stems
  → closing stomata during daylight
→ *resistance to dessication*
  → leaves reduced to spines with photosynthesis carried out by the stem
  → lignified cells supporting the leaf, preventing wilting and reduction in surface area for photosynthesis

**Checkpoint 3**

Using the term 'water potential' explain how more water is lost from a leaf on a dry, windy day than on a humid, still day.

**The jargon**

*Anatomical* features are internal.
*Morphological* features are external.

**Checkpoint 4**

What is meant by the term 'water stress'?

**Checkpoint 5**

Explain how sunken stomata can reduce transpiration.

**Action point**

Draw a labelled diagram of one example of a xerophytic adaptation as seen under the microscope.

**Don't forget**

The rate of transpiration is measured indirectly using an instrument called a potometer.

---

**Exam question** answer: page 50

*Hakea* is a plant which lives in the Australian desert. The diagrams below show sections of one of its leaves as seen through a microscope under low power and high power.

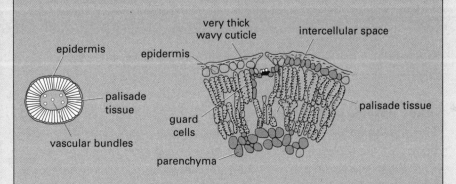

Suggest three ways in which the structure of its leaves helps *Hakea* survive in the hot, dry desert conditions. (5 min)

43

# Transport of water through the plant

The transport of water and minerals takes place in specialized vascular tissue, known as xylem. As you study this topic think of the forces of 'push' at the roots and 'pull' at the leaves.

## Examiner's secrets

The distribution of vascular tissues differs in stems, roots and leaves. You should have some understanding of this from microscope studies of tissues and individual cells.

## Checkpoint 1

How does lignification occur?

## Test yourself

How is the structure of xylem related to its role of water transport?
Draw the four different types of xylem cells in LS and TS.
Consider the evidence supporting the view that xylem carries water up the stem.

## Links

The effect of environmental conditions on transpiration is described on page 43.

## The jargon

*Plasmodesmata* are tiny strands of cytoplasm which link the cytoplasm of one cell to that of the next.

## Checkpoint 2

What is the other function of root hairs?

## Checkpoint 3

Is osmosis an active or passive process?

## Structure of xylem

**Xylem** is made up of four different types of cells: vessels, tracheids, fibres and xylem parenchyma. Vessels and tracheids are the cells involved in conduction. These are dead cells and they form a system of tubes through which water can travel. The cells are dead because lignin has been deposited on the cellulose cell walls rendering them impermeable to water and solutes. These cells also provide mechanical strength and support to the plant.

## The movement of water through the plant

Water enters the plant mainly through the root hairs by osmosis, travels across the **cortex** and into the xylem. It then travels in the xylem up through the stem to the leaves, where most of it evaporates from the internal leaf surface and passes out, as water vapour, into the atmosphere.

The transpiration of water from the leaves draws water across the leaf from the xylem tissue along three pathways:

→ the **apoplast** – through the cell wall and intercellular spaces
→ the **symplast** – through the cytoplasm and plasmodesmata
→ the **vacuolar pathway** – from vacuole to vacuole

The transpiration of water causes a more negative water potential in the mesophyll of the leaf. As water in the xylem is of a higher (less negative) water potential, it moves up the xylem to replace the water lost. This movement of water is known as the transpiration stream. However, water also moves up the xylem by capillarity.

Capillarity has two components:

→ the water molecules have an attraction for each other (cohesion) so when one water molecule moves upwards, others move with it
→ the water molecules are attracted to the sides of the xylem vessels pulling the water upwards (adhesion). As xylem consists of very long, narrow vessels (or 'tubes'), considerable adhesive forces are developed, sufficient to support a considerable mass of water. The forces combine to maintain the column of water in the xylem.

This is known as the cohesion–tension theory and suggests an explanation for the rise of water in the xylem of all plants.

## Water uptake by the roots

The large quantities of water lost through transpiration must be replaced from the soil. The region of greatest uptake is the root hair zone where the surface area of the root is enormously increased by the presence of root hairs on the cells of the piliferous layer.

Most of the water enters the root from the soil down a gradient of water potential. It then passes across the root cortex along two main pathways, the symplast and apoplast pathways, until it reaches the endodermis. Most of the water probably follows the latter pathway, which is the faster of the two. The cells of the endodermis have layers of suberin around them, forming a distinctive band known as the Casparian strip. This prevents the use of the apoplast pathway, so the water must pass through the cytoplasm of the cells in this layer, the only available route to the xylem. There is some evidence that salts may then be actively secreted into the vascular tissue from the endodermal cells. This makes the water potential in the xylem more negative, causing water to be drawn in from the endodermis, so promoting the movement of water into the xylem from the cortex. The water potential gradient produced creates a force known as the root pressure.

## The uptake of minerals ●●●

Generally, minerals are taken up by the root hairs by active transport from the soil solution. Once absorbed the mineral ions may move along the apoplast pathway carried along in solution by the water being pulled up the plant in the transpiration stream. When minerals reach the endodermis the Casparian strip prevents further movement along the cell walls. The ions enter the cytoplasm of the cell, from where they diffuse or are actively transported into the xylem.

**The jargon**

The *Casparian strip* is a thickening of a waxy material, suberin, on the radial walls of the endodermal cells. This impermeable layer diverts the water forcing it to take the symplast pathway.

**Checkpoint 4**

What happens if the stem of a plant is cut near the root during periods of rapid growth?

**Action point**

Construct a diagram of the plant showing the various pathways of water transport and indicate whether the processes involved are active or passive.

**Checkpoint 5**

What are the requirements of the active uptake of minerals?

**Checkpoint 6**

What is the effect of a respiratory inhibitor on mineral uptake?

**Exam questions**                                             answers: page 50

1   The diagram below represents part of a transverse section across a young root. The relative distance across the cortex has been shortened for simplicity.

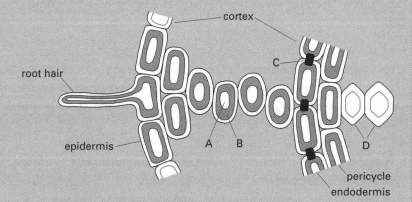

(a) (i)   Name the features labelled A, B, C and D.
    (ii)  State two advantages to the plant of root hairs.
    (iii) What is the function of feature C?
(b) It is believed that water can move across the cortex along the cellulose cell walls. Suggest how this movement is maintained.

(10 min)

# Translocation in plants

The products of photosynthesis are transported in the phloem, away from the site of synthesis (the 'source') in the leaves, to all the other parts of the plant (the 'sink') where they are used for growth or storage. In plants the transport of the soluble organic materials, sucrose and amino acids is known as translocation.

**Action point**

Draw a sieve tube and companion cell.

**Examiner's secrets**

You should be able to list the experimental evidence that phloem is the transport tissue.

## Structure of phloem

Phloem is a living tissue and consists of four types of cells:

→ sieve tubes
→ companion cells
→ phloem fibres
→ phloem parenchyma

The sieve tubes are the only components of phloem obviously adapted for the longitudinal flow of material. They are formed from cells called sieve elements placed end to end. The end walls do not break down but are perforated by pores. These areas are known as sieve plates. Cytoplasmic filaments containing phloem protein extend from one sieve cell to the next through the pores in the sieve plate. The sieve tubes do not possess a nucleus and most of the other cell organelles disintegrate. Each sieve tube element is closely associated with one or more companion cells, which have dense cytoplasm, large centrally placed nuclei, many mitochondria, and are connected to the sieve tube element by plasmodesmata.

**Checkpoint 1**

Which two food materials are transported in the phloem?

**Checkpoint 2**

Explain in terms of water potential the important differences between a 'source' and 'sink' region of the plant.

**Checkpoint 3**

What parts of the plant require sugars?

## Transport in the phloem

The observed rate of flow is much too rapid for diffusion to be the cause. The main theory put forward to explain the transport of organic solutes is known as the mass flow hypothesis (put forward in 1937 by Ernst Munch). This suggests that there is a passive mass flow of sugars from the phloem of the leaf where there is the highest concentration (the source) to other areas where there is a lower concentration (the sink).

→ At the source, phloem companion cells actively take up sugars and pass it to the sieve tubes while the reverse process occurs at the sink. Transfer cells, which occur in the mesophyll, are found to be associated with the phloem tissue and are thought to be involved in this process.
→ This causes a concentration gradient from source to sink.
→ Water is drawn in by osmosis, setting up a hydrostatic pressure resulting in movement of sugar to the sink.
→ At the sink the sugar concentration decreases, as it is used for metabolic processes, so water passes out into the tissues.

→ The continual input of sugars and water at the top of the system and their removal at the bottom creates a pressure gradient that maintains the downward flow of fluid in the sieve tubes.

There are drawbacks to the theory.

→ It does not explain the existence of the sieve plates, which seemingly act as a series of barriers impeding flow.
→ Sugars and amino acids have been observed to move at different rates and in different directions in the same tissue.

Other hypotheses have been proposed. These include, electro-osmosis, surface spreading and cytoplasmic streaming along the protein filaments.

**Action point**

Review the evidence for and against the mass flow hypothesis.

**Watch out!**

Don't confuse transpiration and translocation.

## Exam questions

answers: page 50

1  (a) With the aid of labelled diagrams, describe the structures involved in the translocation of organic solutes between different parts in a flowering plant. (15 min)
   (b) Explain one experiment which has been performed that has enhanced our knowledge of translocation in flowering plants. (8 min)

2  (a) (i)  Complete the table below which compares xylem and phloem.

| Feature | Xylem | Phloem |
|---|---|---|
| Name of conducting cells | | |
| Direction in which materials are transported | | |
| One possible mechanism by which materials are transported | | |

   (ii) Briefly describe *one* way of demonstrating movement of fluid in phloem.
   (b) The table below shows the concentration of two chemicals to be found in the phloem and xylem sap of the white lupin, *Lupinus albus*.

| Feature | Xylem (mg $l^{-1}$) | Phloem (mg $l^{-1}$) |
|---|---|---|
| Sucrose | not detected | 154 000 |
| Magnesium | 27 | 85 |

Suggest why the concentration of sucrose and magnesium differs in xylem and phloem.

(15 min)

**Don't forget**

Radioactive labelling techniques have been used to investigate the mechanism of translocation. A plant is able to use radioactive carbon dioxide in photosynthesis to produce labelled sugars.

# Answers
## Exchange and transport

## Movement of molecules

### Checkpoints

1 An increase in temperature results in an increase in rate since there is an increase in molecular energy and therefore movement.
2 Most of the membrane consists of lipophilic tails, which tend to repel ions and polar molecules. These can only diffuse across slowly whereas molecules with a high solubility in fat are taken up rapidly.
3 Shrink/shrivel.
4 A to B.
5 Solutions of equal concentration are called isotonic. A more dilute solution is called hypotonic. The more concentrated solution is called hypertonic.
6 Increases it since the rate of respiration is increased.

### Exam questions

1 (a)

| Active transport | Facilitated diffusion | Osmosis |
|---|---|---|
| ✓ | | |
| ✓ | ✓ | |
| | ✓ | ✓ |

(b) (i) Cell is plasmolyzed/cytoplasm shrunk from cell wall.
(ii) Cell wall is permeable (to sugar as well as water)/same solution on both sides of cell wall.

## Human digestion

### Checkpoints

1 Thorough mixing of food and gastric juice.
2 An exocrine gland has a duct through which the secretions pass. In an endocrine gland blood flows through it and receives the secretions directly.
3 Both, since it produces insulin (endocrine) and digestive enzymes (exocrine).
4 To prevent autodigestion of the cells of the gastric glands in which they are formed.
5 Mouth – alkaline; stomach – very acid; duodenum – alkaline.
6 Water, minerals, vitamins.
7 Glucose, amino acids, fatty acids.

### Exam question

- General diagram of tissue layers showing outer layer of longitudinal and inner layer of circular muscle, inner mucosa, separated from muscle by submucosa.
- A labelled diagram of the gut wall showing villi, microvilli, epithelium, columnar cells, goblet cells, crypts, lacteals, capillaries, Brunner's glands (in duodenum only).
- Function of muscle, movement of food along gut by peristalsis.
- Goblet cells producing mucus.
- Brunner's glands producing alkaline mucus to aid neutralization of acid from stomach.
- Cells in crypts of Lieberkuhn secrete enzymes.
- Length of gut important to allow time for enzyme action and absorption.
- The villi and microvilli increase surface area for absorption by epithelial cells.
- Further enhanced by microvilli on cells.
- In each villus capillaries 'pick up' simple sugars and amino acids which are carried away by the hepatic portal vein.
- Glycerol and fatty acids are removed by lacteals and are carried into the main lymphatic system.

> **Examiner's secrets**
>
> This question is specifically about the small intestine so don't be tempted to waste valuable time describing mouth and stomach. Be precise about which secretions are produced by the various glands.

## Gas exchange in plants and simple organisms

### Checkpoints

1 They would rapidly dessicate on land.
2 Being flattened they have a large surface area/also sluggish and have a low metabolic rate.
3 In daylight photosynthesis masks the respiratory production of carbon dioxide because the respiratory carbon dioxide is immediately reused in the chloroplasts.
4 Epidermal cells.
5 Humidity, wind, temperature.
6 Mitochondria.

### Exam question

The answer must include a diagram for full marks to be awarded. It should be clear and accurate with at least six labelled structures, including guard cells, thick inner wall and thin outer wall, nuclei, chloroplasts, cell vacuole.
The diagram should show a bending of the inner wall and opening of aperture in a turgid cell.
The mechanism is as described in the text.
The functions are to allow gaseous exchange and to prevent loss of water by closure in water-stressed plants.

## Gas exchange in mammals

### Checkpoints

1 Haemoglobin.
2 Rings of cartilage.
3 Thin, moist, permeable, large surface area.
4 Ventilation rate = number of breaths per minute (16) × tidal volume (16 × 0.45 = 7.2).
5 Increase in frequency and depth.
6 Allows the study of oxygen consumption in humans and can be used to record depth and frequency of breathing.
7 After maximum expiration this is the amount of air remaining in the lungs.

## Exam question

The answer must include a diagram for full marks to be awarded with at least six labelled structures. Include the following words in annotated labelling: pharynx, larynx, epiglottis, trachea, bronchi, cilia and mucus-secreting cells, bronchioles, alveoli, network of capillaries.

## Transport in mammals: blood

### Checkpoints

1 Liver.
2 Increased surface area and shorter diffusion path to carry oxygen.
3 It picks up oxygen more readily but is less ready to release it.
4 The haemoglobin has a low affinity for oxygen and therefore unloads it quickly to the tissues.
5 To ensure that the overall charge remains neutral.

### Exam questions

1 (a) Carbon dioxide.
   (b) Lungs.
   (c) Chloride shift as described in the text.
   (d) It preserves the ionic balance on either side of the membrane.

## Transport in mammals: vessels

### Checkpoints

1 To withstand and maintain the pressure of blood when it is pumped away from the heart.
2 To slow blood flow to allow substances time to diffuse.
3 Contraction of skeletal muscles/inspiration.
4 Help prevent infection/lymphocytes produced in lymph nodes/tissue fluid bathes cells.
5 Tissue fluid lacks proteins (and blood cells).

### Exam questions

1 (a) (i) A, artery; B, arterioles; C, capillaries; D, venule; E, veins.
      (ii) 1: B. 2: C.
   (b) (i) Contractions of left ventricle cause blood to surge into the artery; relaxation of the left ventricle results in a fall in pressure; the amplitude change is due to the elasticity of the vessels which reduces with distance from the heart.
      (ii) An increase in the total cross-sectional area of the vessels.
   (c) In the capillaries blood moves slowly because of the total cross-sectional area, maximizing time for the exchange of materials between the blood and tissues.
   (d) The massaging action of skeletal muscles; suction effects of the heart.
2 (a) Include any six of the following.
      • Water: a solvent, maintains osmotic balance, conducts heat away from the core to the surface.

• Proteins: act as buffers or involved in blood clotting.
• Hormones: a mention of coordination.
• Antibodies: defence.
• Digested food products: carried from gut to cells.
• Urea: excretory product of cells.
• Metal ions, e.g. $Na^+$: osmotic regulation.
• $HCO_3^-$: pH regulation/carbon dioxide transport.

(b)
• Capillary lining consists of a single layer of cells (endothelium).
• Ventricular contraction exerts an outward pressure (hydrostatic pressure) on the capillary contents.
• The blood proteins exert a more negative water potential.
• Hydrostatic pressure is greater than the water potential at the arterial end and water, ions and small molecules are forced out.
• Loss of water means that water potential is greater than hydrostatic pressure at the venous end and water is drawn back, leaving small molecules concentrated in tissue fluid.

## Transport in mammals: heart

### Checkpoints

1 Pulmonary.
2 Pumps blood all round the body.
3 Prevent valves from turning inside out.
4 Tricuspid valves shut.
5 100 000 times.

### Exam questions

1 (a) A is the sympathetic/accelerator nerve. It increases the heart rate. B is the sinoatrial node or pacemaker. It regulates the heart rate/initiates heartbeat. C is the bicuspid valve and prevents back-flow of blood into the left atrium.
   (b) (i) Medulla oblongata.
      (ii) Increases stroke volume/increases heart rate; increases blood flow to the lungs (cardiac output); increasing the removal of carbon dioxide/restores partial pressure of carbon dioxide.
   (c) It reduces the volume of blood circulating/reduces blood pressure/fluid in the lungs.

## Transport in plants

### Checkpoints

1 In the stem the vascular tissue is found on the periphery; in the root it is found in the centre.
   In the stem xylem and phloem are on the same radius; in the root they are on alternate radii.
2 In the stem the peripheral position counteracts a bending strain; in the leaf the network of veins resist the tearing effect of wind; in the root the central position resists pulling by wind.

3 The sub-stomatal chamber has a high water potential as the walls of the spongy mesophyll cells are saturated with water. The water evaporates from the walls and moves down a gradient of potential from the plant to the atmosphere.

4 If a plant loses more water through transpiration than it can take up into its roots.

5 The water potential gradient is reduced as the stomata are enclosed within a 'microclimate' with epidermal hairs helping to maintain high humidity.

### Exam question

A reference should be made to the following methods of reducing evaporation: rolled shape giving reduced surface area; thick waxy cuticle impervious to water; stomata sunken in pits with chamber above stoma becoming saturated with water.

## Transport of water through the plant

### Checkpoints

1 The deposition of lignin between cellulose microfibrils in the cell wall.

2 Anchorage.

3 Passive.

4 Sap exudes due to root pressure.

5 Oxygen and glucose providing energy by respiration in the mitochondrion.

6 Mineral uptake reduced or stopped as it is energy requiring.

### Exam questions

1 (a) (i)  A, vacuole; B, cytoplasm; C, Casparian strip; D, xylem.

(ii)  Increase surface area for uptake of minerals and water; anchorage.

(iii)  It prevents water moving along the apoplast pathway and so means everything must go through the symplast pathway.

(b)  There is a mass flow of water pulled by evaporation from the leaves and water molecules are held together by cohesive forces.

## Translocation in plants

### Checkpoints

1 Sucrose and amino acids.

2 Sieve tubes in a leaf are a 'source' region for pressure flow. Phloem in leaves contains a high concentration of sucrose and therefore will have a low (more negative) water potential. Water flows into the phloem from the xylem and a high hydrostatic pressure is created. Roots are a 'sink' region. In the cells of a root sugars are turned to starch. There will be a relatively high (less negative) water potential. Water flows into neighbouring cells by osmosis and a low hydrostatic pressure develops.

3 Growing points, food storage organs.

### Exam questions

1 (a) The diagrams needed are of phloem tissue, a sieve tube and a companion cell; a diagram showing the relationship of the vascular tissue with transfer cells could be helpful also. Give a written description of these cells and their structures as well.

(b) One of the following experiments should be described: removal of a ring of bark, using radioactive tracers **or** using aphid mouthparts.

2 (a) (i)  Xylem: tracheids or vessels transport materials upwards from roots to leaves by transpiration pull (stream) **or** cohesion–tension or root pressure. Phloem: sieve tubes (elements) transport materials up and down by mass flow **or** cytoplasmic streaming **or** transcellular strands **or** electro-osmosis.

(ii)  Allow aphid stylet to penetrate phloem, remove body, fluid oozes out **or** supply plant with radioactive-labelled $CO_2$ and follow sucrose using autoradiography **or** describe a ringing experiment.

(b) Sucrose: synthesized in the leaves by photosynthesis; actively loaded; transported to sinks in the phloem which may be above or below (taken to other parts of the plant).
Magnesium: imported from the soil to growing points or leaves in the xylem where some is used to manufacture chlorophyll; active transfer of magnesium from xylem to phloem for redistribution.

# Energy and the environment

The study of the flow of energy through the ecosystem is known as ecological energetics. The source of energy in an ecosystem energy is light, and this is converted to chemical energy by the process of photosynthesis. The energy then passes along the food chain, at each stage of which some of the energy is lost as heat. Unlike energy, minerals such as carbon and nitrogen can be continuously recycled. Applied ecology requires information about the size of plant and animal populations so that humans can regulate and manage populations of organisms for their own use. The increasing pressures placed upon the natural environment have created problems and humans have come to understand the need for more careful conservation of resources.

## Exam themes

What is meant by an ecosystem?
The efficiency of energy transfer between trophic levels
The advantages and disadvantages of the various forms of ecological pyramids
The cycling of nitrogen within an ecosystem
Factors affecting population growth
Predator–prey relationships
The advantages and disadvantages of organic farming
The implications of intensive food production in terms of the effect of farm waste, land reclamation and hedgerow destruction
The human exploitation of resources and the need for conservation
Methods of pest control
The impact of agriculture on the environment

## Topic checklist

| ○ AS　● A2 | AQA/A | AQA/B | EDEXCEL | OCR | WJEC |
|---|---|---|---|---|---|
| Autotrophic and heterotrophic nutrition | | | ○ ● | ○ | ○ |
| Energy flow in the ecosystem | ● | ● | ○ | ○ | ○ |
| Food chains and food webs | ● | ● | ○ | ○ | ○ |
| Ecological pyramids | ● | ● | ○ | ● | ○ |
| Recycling of nutrients | ● | ● | ○ | ○ | ○ |
| Population growth and control | ● | ● | ● | ● | ○ |
| Resource management and human influence | ○ | ● | ○ | ○ | ○ |

# Autotrophic and heterotrophic nutrition

Most autotrophic organisms use the simple organic materials, carbon dioxide and water, to manufacture energy-containing complex organic compounds, whereas heterotrophic organisms consume complex organic food material.

## Autotrophic nutrition

There are two types of autotrophic nutrition.

### Photosynthesis

This is the process by which green plants, algae and certain types of bacteria build up complex organic molecules from carbon dioxide, water and mineral ions. The source of energy for this process comes from sunlight which is absorbed by chlorophyll and related pigments.

### Chemosynthesis

This is the process by which a few bacteria can perform similar synthesis of organic compounds using energy derived from special methods of respiration.

## Heterotrophic nutrition

Heterotrophic organisms consume complex organic food material. There are a number of different forms of heterotrophic nutrition.

### Holozoic

This involves taking in complex organic molecules, breaking them down by digestion, absorption into the body tissues from the digestive system and, finally, utilization of the absorbed products of digestion in the body cells. Animals that feed solely on plant material are termed **herbivores**, those that feed on other animals are **carnivores**, and the **omnivores** have a mixed diet.

### Saprobiontic

This is also known as saprotrophic or saprophytic. It involves the absorption of complex organic food from the bodies of decaying organisms. Some bacteria and fungi feed in this way. They secrete enzymes on the food substrate and then absorb the products of this extracellular digestion. The activities of these organisms are important in the decomposition of leaf litter and the recycling of valuable nutrients.

### Parasitic

A parasite is an organism that lives in or on another living organism, the host. The parasite derives all its nutrition from the host, whereas the host does not gain any benefit from the association and is often harmed to some degree. Some parasites live on the outside of the host, the **ectoparasites**, e.g. leech. Others live entirely within the body of the host and are termed **endoparasites**, e.g. Plasmodium, the malarial

---

**Links**

Photosynthesis is looked at in detail on pages 76 and 77.

**Checkpoint 1**

List two differences between autotrophic and heterotrophic nutrition.

**Links**

Food chains are dealt with on pages 56 and 57.

**Checkpoint 2**

State one harmful effect of saprobionts.

**Examiner's secrets**

Examiners are impressed when candidates name specific examples.

**Checkpoint 3**

Why do most parasites produce very large numbers of eggs or spores?

parasite. Parasites are considered to be very highly specialized organisms and show considerable adaptations to their mode of life.

## Mutualistic

This is also known as symbiosis. It also involves a close association between members of two species, but in this case both derive some benefit from the relationship, e.g. the digestion of cellulose by microorganisms in the gut of a herbivore.

**Checkpoint 4**

Give another example of mutualism.

**Checkpoint 5**

Which of the following types of organisms can live independently of their food supply: holozoic, saprobiontic, parasitic?

---

**Exam questions**                                                    answers: page 66

1   The terms 'autotrophic', 'heterotrophic' and 'parasitic' are used to describe the ways in which different organisms obtain their nutrients. The table shows a list of organisms. Complete the table by ticking the box or boxes which apply to each organism. It may be necessary to tick more than one box to fully describe the way in which nutrients are obtained. (8 min)

| Organism | Autotroph | Heterotroph | Saprophyte | Parasite |
|---|---|---|---|---|
| Apple tree | | | | |
| HIV | | | | |
| Dog | | | | |
| Nitrogen-fixing bacteria | | | | |
| *Penicillium* | | | | |
| Tapeworm | | | | |

2   Using the diagram below explain why
   (a) (i)   spore cases are found on long 'stalks'
       (ii)  the 'bodies' (mycelium) of the fungus is always found in the surface layer of the substrate and not deep inside the tissues
       (iii) the mycelium covers a large area
   (b) What is meant by extracellular digestion?

(7 min)

**Examiner's secrets**

Don't tick too many boxes in the hope that you'll have covered the right ones. Examiners will penalize you for this.

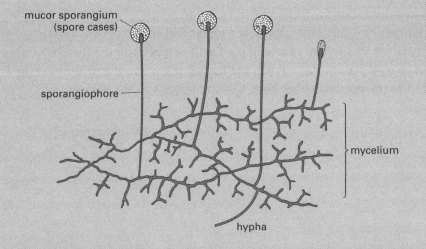

mucor sporangium (spore cases)

sporangiophore

mycelium

hypha

**The jargon**

A *mycelium* is a collection of hyphae.

# Energy flow in the ecosystem

An ecosystem is a community of organisms that, together with their physical environment, form a self-perpetuating ecological unit. The study of the flow of energy through the ecosystem is known as ecological energetics.

## Components of an ecosystem

**Checkpoint 1**

Which terms describe the following features of a seashore: all limpets on the seashore; temperature variations in a rock pool?

→ A *population* is a group of organism of a single species occupying a particular area. In this area there will also be other populations, forming a community.

→ A *habitat* is the particular area occupied by a population. It has biotic and abiotic features that separate it from other habitats. The biotic features are the sum total of the organisms within the habitat and their interactions. Abiotic features include different types:

→ **edaphic** features relate to the soil and include all its physical and chemical characteristics;

→ **climatic** features include light, temperature, moisture, salinity and, particularly, the stability or variability of these.

→ *Microhabitats* are small localities within a habitat, each with its own particular conditions.

**Checkpoint 2**

What is the main difference between a niche and a habitat?

→ *Ecological niche* is the place of each species in an ecosytem. This is not only the physical space that it occupies, but the role which it carries out within the community and its interrelationships with other species as well. In the long term, two species cannot occupy the same niche in a specific habitat.

## Energy source

**Checkpoint 3**

What is the ultimate fate of most energy on the earth?

The ultimate source of energy for ecosystems is the sun, from which energy is released in the form of electromagnetic waves. Only a small part of the total amount of energy reaching the Earth's atmosphere enters ecosystems. Also, the quantity absorbed by plants varies considerably at different latitudes. Of the energy entering a plant only 1–5% is utilized by the plant; the rest is lost, partly by reflection and partly by the evaporation of water.

## Energy flow

The energy flowing from one organism to another in the food chain is not recycled but eventually lost as heat and so it must constantly be replaced by sunlight reaching the Earth. If it is assumed that 100 units of energy per unit time reach the leaves of a plant, the diagram at the top of the opposite page shows what happens to the energy.

fate of solar energy reaching the leaf of a crop plant (numbers are percentages of the total solar energy falling on the leaf)

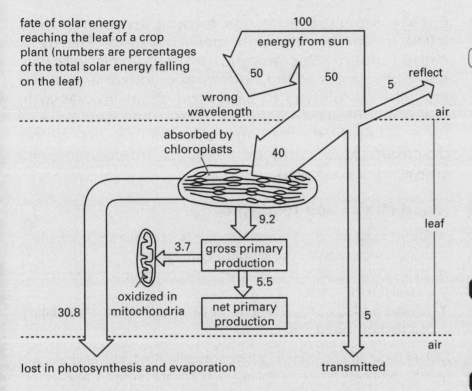

lost in photosynthesis and evaporation                    transmitted

Gross primary productivity (GPP) is the rate at which products are formed. A substantial amount of gross production is respired by the plant.

GPP – respiration = net production

The rate at which the products of photosynthesis accumulate is known as net primary productivity (NPP). Secondary productivity is the rate at which consumers accumulate energy in the form of cells or tissues.

**Checkpoint 4**

If the NPP of grass = 21 135 kJ m$^{-2}$ year$^{-1}$ and a bullock eats 3 056 kJ m$^{-2}$ year$^{-1}$ what happens to the rest of the energy?

**Checkpoint 5**

When food is in short supply explain why more vegetarians can be fed than those who must have meat.

**Checkpoint 6**

Why do carnivores have a much higher secondary productivity than herbivores?

**Exam questions**                                    answers: page 66

1   The diagram below shows the quantity of energy flowing through a food chain in a terrestrial ecosystem. The figures given are kJ m$^{-2}$ year$^{-1}$.

incident sunlight $3 \times 10^6$

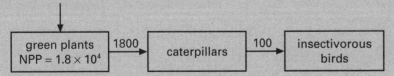

(a) Calculate the percentage of the incident energy which becomes available as the net primary production (NPP) of green plants. Show your working.

(b) Give two reasons why not all the energy of the incident sunlight is incorporated into the biomass of green plants.

(c) Using the information shown in the diagram, explain why the biomass of insectivorous birds is usually very much less than the biomass of caterpillars.

(8 min)

2   Comment on the flow of energy through ecosystems and discuss the various ways in which human activity can influence this flow at all levels in terrestrial ecosystems. (15 min)

**Examiner's secrets**

Marks are allocated for showing the working in calculations. Don't do everything on the calculator giving only the answer.

# Food chains and food webs

Energy is passed from one feeding level (or trophic level) to another through the ecosystem. Energy is passed along a hierarchy of trophic levels with primary producers at the bottom and consumers at the top. This is referred to as a food chain. However, in practice, organisms feed off more than one type of food. Therefore, feeding relationships are better described as a complex series of interconnections known as a food web.

## Food chains and food webs

→ Green plants which manufacture sugars from simple raw materials using solar energy are called **primary producers**.
→ Animals feeding on these plants are called herbivores or **primary consumers**.
→ Animals feeding on these animals are called carnivores or **secondary** and **tertiary consumers**.

Each of these groups forms a feeding or trophic level, with energy passing from each level to a higher one as material is eaten. At each level energy is lost through respiration and in waste products, so the amount of energy is reduced.

The sequence from plant to herbivore to carnivore is a food chain and is the route by which energy passes between trophic levels. It is the loss of energy at each level which limits the length of a **food chain** so the number of links in a chain is normally limited to four or five.

On the death of producers and consumers, some energy remains locked up in the organic compounds of which they are made. **Decomposers** feed as saprobionts and contribute to the recycling of nutrients (see diagram below).

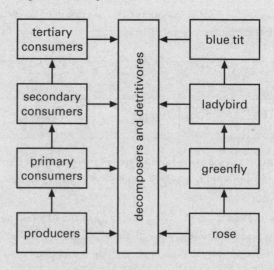

Single food chains rarely exist and a **food web** is a more realistic representation of feeding relationships. This is because most primary consumers feed on more than one kind of autotroph and most secondary consumers rely on more than one type of prey. The more varied the organisms in an environment, the more complex the food web.

**Checkpoint 1**

What is the ultimate source of energy of a food web?

**Watch out!**

Candidates often perform poorly in ecology because they do not learn the definitions of the terms used.

**Checkpoint 2**

What is the difference between a decomposer and a detritivore?

**Links**

Recycling of nutrients is looked at on pages 60 and 61.

**Checkpoint 3**

What is the main difference between a food chain and a food web?

**Watch out!**

A single species may form part of many different chains, not always occupying the same trophic level in each chain.

## Trophic efficiency

This is the percentage of energy at one trophic level which is incorporated into the next trophic level. The rate at which energy passes into the animals at each trophic level is about 10% of that entering the previous level. This is called the gross ecological efficiency. This value differs from one ecosystem to another, with some of the highest values, around 40%, occurring in oceanic food chains. Some of the lowest values, around 1%, are found in ecosystems where most of the animals are birds or mammals.

**Checkpoint 4**

What happens to the other 90% of the energy?

**Checkpoint 5**

Why is the trophic efficiency of small mammals generally less than 1%?

---

**Exam questions**                                    answers: page 66

1   The diagram below shows part of a food web in a freshwater pond. Name different organisms from the web which proved examples of:

   (a) an omnivore

   (b) a producer

   (c) a secondary consumer

   (d) a tertiary consumer

(4 min)

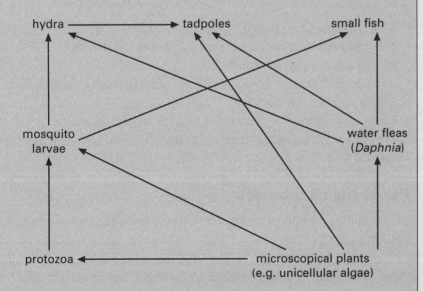

2   (a) Plants in a grassland ecosystem assimilate about 1% of the sun's radiant energy that falls on them. Approximately 80% of the plants' assimilated energy is available to the next trophic level.

      (i)  Give two reasons why plants assimilate so little of the sun's energy.

      (ii) What happens to the energy assimilated by plants that is not available to the next trophic level?

   (b) Explain why the number of links in food chains does not often exceed five.

(10 min)

**Action point**

For a habitat other than a freshwater pond, construct a food web by interconnecting all the organisms using arrows. Don't forget that the arrow heads indicate the direction of energy flow.

**Watch out!**

Be aware that the same organism can occupy different trophic levels in a food web. For example, in the diagram opposite a tadpole is a primary consumer when it feeds on algae but a secondary consumer when it feeds on water fleas.

# Ecological pyramids

The number of organisms, their biomass or the amount of energy contained in each trophic level can be represented in diagrams with a bar for each level. These are known as pyramids. You should be aware of the advantages and disadvantages of each of the three types of pyramids used in ecology.

## Pyramids of numbers

If a bar diagram is drawn to indicate the relative numbers of individuals at each trophic level in a food chain, a diagram similar to that below is produced.

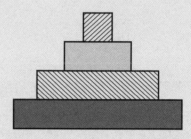

The overall shape is roughly that of a pyramid with the number of organisms at each level usually decreasing. This is because the loss of energy through the chain means that the higher levels can support fewer numbers and also the animals at the top tend to be larger anyway.

There are drawbacks to the use of pyramids of numbers.

→ All organisms are counted as equal, regardless of size.
→ This can lead to inverted pyramids, e.g. a single oak tree supporting millions of aphids.
→ Juvenile or immature forms may have different energy and diet requirements to the adult.
→ The numbers of some organisms may be so great that it is impossible to represent them accurately to the same scale as other species in the food chain.

## Pyramids of biomass

The difficulties mentioned are partly overcome by the use of a pyramid of **biomass**. This reflects the decrease in biomass at each trophic level in a food chain. The biomass is the total dry mass (in kg m$^{-2}$) of living matter, so it reflects both the numbers of organisms at each trophic level and also their size. It too has its drawbacks.

→ Since it is impossible to measure exactly the biomass of all individuals in a population, a small sample is taken.
→ If the producers are short-lived, at any one time the mass of producers will be smaller than that of the primary consumers. This is called the **standing crop**. However, over a period of time, the biomass of the producers would be greater.

**Check the net**

You'll find information on ecological pyramids at www.sturgeon.ab.ca/rw/Pyramids/ecopyra.html

**Checkpoint 1**

Label the trophic levels in the diagram.

**Checkpoint 2**

How is biomass measured?

**Checkpoint 3**

What is the disadvantage of taking a sample?

**Checkpoint 4**

Explain why zooplankton (microscopic organisms) may have a greater increase in biomass than a tree in the same period of time?

# Pyramids of energy ●●●

This represents the total energy requirement of each successive trophic level in a food chain. As material passes up through the food chain energy is lost in respiration as heat, and in excretion, so the size of the bars decreases sharply. They are usually constructed for the energy utilized by the different feeding types in a unit area over a set period of time. The use of a set period of time means that an energy pyramid overcomes the problems which arise when ecosystems are compared simply by counting or measuring the standing crop of organisms. However, obtaining the data can be complex and difficult.

**Checkpoint 5**

How is the energy of an ecosystem measured?

**The jargon**

*Standing crop* describes the amount of living material present at a given instant in time. The energy pyramid introduces a time factor and represents the energy flow per unit time (say, one year). It therefore gives the best overall view of the community.
Units are kJ/m$^2$/y$^{-1}$

**Exam questions**                                   answers: page 67

1  (a) (i)  Explain what is meant by a pyramid of numbers.
       (ii)  Explain what is meant by a pyramid of biomass.
       (iii) Which pyramid is more useful when considering the productivity of an ecosystem? Explain your answer.
   (b) The diagram below shows the pyramid of biomass in a marine system measured at a given time.

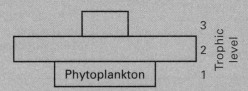

       (i)  Describe how this pyramid of biomass differs from that you would normally expect to find in a terrestrial ecosystem.
       (ii) Suggest two reasons for this difference.
   (c) The table below shows the energy taken up by trophic levels 1, 2 and 3 of a food chain in a marine system compared with a terrestrial ecosystem.

|  | Energy taken in (J m$^{-2}$ day$^{-1}$) | |
| --- | --- | --- |
| Trophic level | Marine ecosystem | Terrestrial ecosystem |
| 1 | 8 400 | 12 500 |
| 2 | 25.5 | 8 |
| 3 | 8 | 1 |

       (i)  The energy taken up from sunlight by trophic level 1 is only a small proportion of the total light energy falling on plants. Explain why.
       (ii) Calculate the percentage of light energy taken up by trophic level 1 which is available to trophic level 3 in a marine ecosystem. Show your working.

                                                    (18 min)

**Examiner's secrets**

Be able to discuss the advantages and disadvantages of the different types of pyramid.

# Recycling of nutrients

Minerals, such as carbon and nitrogen, can be continuously recycled and so used again and again by organisms. Microorganisms play an important role in the process of decay, releasing compounds of these elements from the bodies of dead organisms.

**Checkpoint 1**

Explain briefly how coal was formed.

**Action point**

The combustion of fossil fuels and the burning of enormous quantities of wood removed by deforestation has resulted in an increase in the concentration of carbon dioxide in the atmosphere. Plan an essay that considers the effects of these human activities.

**Checkpoint 2**

What climatic change has taken place due to the removal of trees in tropical rain forests?

**Don't forget**

Nitrogen is found in all amino acids which make up proteins. Nitrogen is available to plants only in the form of ammonium ions ($NH_4^+$) and nitrate ($NO_3^-$) ions which are taken up by the roots.

**Examiner's secrets**

Candidates often find the nitrogen cycle confusing, so make sure you are clear about it.

**Checkpoint 3**

Is the conversion of ammonium to nitrite an oxidation or a reduction process?

## Carbon cycle

Carbon dioxide is added to the air by the respiration of animals, plants and microorganisms, and by the combustion of fossil fuels. Photosynthesis takes place on so great a scale that it reuses on a daily basis almost as much carbon dioxide as is released into the atmosphere. This is the basis of the carbon cycle. The production of carbohydrates, proteins and fats contributes to plant growth and subsequently to animal growth through complex food webs. The dead remains of both plants and animals are then acted upon by saprobionts in the soil, which ultimately release gaseous carbon dioxide back to the atmosphere.

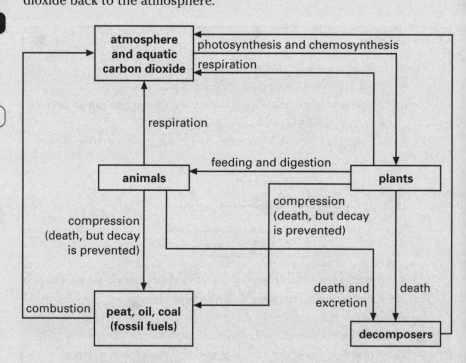

## Nitrogen cycle

The nitrogen cycle is the flow of organic and inorganic nitrogen within an ecosystem where there is an interchange between nitrogenous compounds and atmospheric nitrogen. The main processes involved are as follows.

→ *Putrefaction*: decay processes convert organic nitrogen into ammonia. Various bacteria and fungi carry out this process.
→ *Nitrification*: the ammonium ions formed in putrefaction are converted by nitrification via nitrites to nitrates, the main absorbable form of nitrogen. Various bacteria are involved,

e.g. ammonium is converted to nitrite by *Nitrosomonas* and nitrite to nitrate by *Nitrobacter*.

→ *Nitrogen fixation*: atmospheric nitrogen can be converted to nitrogen compounds by nitrogen fixation, e.g. by free-living bacteria such as *Azotobacter*. In legumes (peas and beans) *Rhizobium* bacteria live in nodules in the roots and are able to fix nitrogen producing $NH_4^+$ ions which enables the plant to make amino acids. In return some carbohydrates are utilized by the bacteria.

→ *Denitrification*: nitrogen is lost from ecosystems by denitrification. This is a particular problem in waterlogged soils with anaerobic conditions where anaerobic bacteria such as *Pseudomonas* can reduce nitrates to molecular nitrogen.

**Checkpoint 4**

Why do denitrifying bacteria thrive in waterlogged soils?

**Checkpoint 5**

What is the link between deforestation and the nitrogen cycle?

**Watch out!**

You should know the specific names of the microbes *Nitrosomonas*, *Nitrobacter* and *Rhizobium*.

**Don't forget**

Decomposition accounts for most of the conversion of organic material from all trophic levels into inorganic compounds that can be recycled.

**Exam questions** answers: page 67

1 The diagram below shows some of the processes involved in the cycling of some nitrogen compounds.

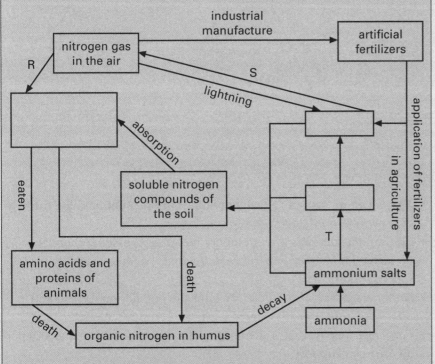

(a) Name the processes labelled R, S and T on the diagram.

(b) Complete the diagram by inserting suitable labels in the empty boxes.

(7 min)

2 Give a biological explanation of why a farmer:

(a) plants a crop of clover after harvesting a cereal crop

(b) should control the amount of nitrogenous material draining into a river

(c) applies organic manure instead of inorganic fertilizers

(8 min)

3 (a) Give a full account of the ways in which nitrogen is exchanged between organisms and their environment.

(b) Discuss the various ways in which human activities influence the nitrogen cycle.

(22 min)

**Examiner's secrets**

Pure recall of the nitrogen cycle is insufficient. Be prepared to answer questions about any part of the cycle.

# Population growth and control

Organisms live as part of populations and communities. You should understand how factors can limit the growth of a population and how the balance between birth rate and death rate is maintained. It is possible to study the development of a community over time.

## Population growth ●●●

A typical S-shaped curve of population growth traces the increase in number of a species moving into a new geographical area (see graph below).

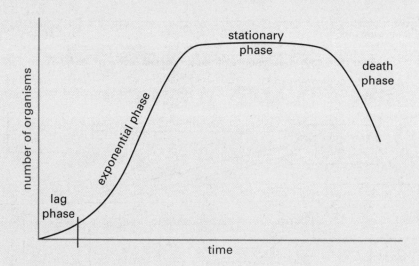

→ The *lag phase*: as only a few individuals are present initially, the rate of growth is very slow.
→ The *exponential phase*: as numbers increase, providing there is no factor limiting growth, more individuals become available for reproduction.
→ *Stationary phase*: certain factors limit the population growth and birth rate equals death rate.
→ *Carrying capacity*: this is the limit to the number of individuals that an area can support.

The factors that limit the growth of a particular population are collectively called the **environmental resistance**. Such factors include available food, predation, overcrowding, disease, accumulation of toxic waste.

## Density dependence ●●●

Some factors are density dependent, i.e. their effect increases as the density of the population increases, e.g. accumulation of toxic waste, disease, parasitism and sometimes food supply.

Other factors are density independent, i.e. their effect does not depend on the population density – the effect is the same regardless of the size of the population. It is usually due to a sudden or violent change in an abiotic factor, e.g. freezing, flood or fires.

**Checkpoint 1**

Explain the reason for the lag phase when yeast is introduced into a solution of glucose.

**Action point**

Draw a line on the graph above to show the carrying capacity of the population.

## Regulation of population size ●●●

The size of a population is regulated by the balance between the birth rate and the death rate. However, there are other influences on the size of a population.

→ *Immigration* occurs when individuals join a population from neighbouring populations.
→ *Emigration* occurs when individuals depart from a population.
→ *Competition* occurs when individuals of a species in a population are continually competing with each other. Plants compete for space, light and mineral ions; animals compete for food, shelter and a mate.

There are two forms of competition:

→ *intraspecific competition* occurs between individuals of the same species
→ *interspecific competition* occurs between individuals of different species

## Predation ●●●

A predator is an organism that feeds on living species. Predators are normally larger than their prey and tend to kill before they eat. The abundance of prey is a factor limiting the numbers of the predator. Within a food chain, a predator–prey relationship causes both populations to oscillate and these oscillations are regulated by the process known as negative feedback.

## Community and succession ●●●

In any area, over time, new organisms replace existing ones until a stable state is reached. Bare rock is colonized by algae and lichens, forming a **pioneer** community. The accumulation of dead and decomposing organic material and the weathering of the rock leads to the formation of a primitive soil in which the higher plants can grow. Mosses and ferns appear and as the soil develops, grasses, shrubs and trees appear. This constitutes a succession and eventually leads to a **climax** community. On land this is usually deciduous woodland with one dominant or several co-dominant species of trees. Human interference produces an unnatural climax, a **plagioclimax**, such as chalk grassland or lowland heath.

**Checkpoint 2**

How can two species living in the same habitat avoid competition?

**Checkpoint 3**

What type of competition exists between a cormorant and a seagull?

**Speed learning**

It's very easy to confuse *intraspecific* and *interspecific* competition. Think of international sports matches **between different countries** – *interspecific competition* occurs **between** individuals of **different species**!

**Links**

Specific predator–prey relationships in biological control are looked at on page 64.

**Checkpoint 4**

What is the value of a negative feedback system in population control?

**The jargon**

*Deciduous* trees are those that lose their leaves in winter.

**Exam question**                                   answer: page 67

Describe what you understand by an ecological climax. (5 min)

# Resource management and human influences

As the size of the human population increased, a more efficient agricultural system had to be developed in order to support it. The development of monocultures and the use of fertilizers, pesticides and better crops were significant factors in the process. However, humans have come to learn that progress is not without its associated problems, and as a result of the increase in pressures placed upon the natural environment there has been a growing interest in conservation.

## Monoculture ●●●

Monoculture is the simultaneous growth of large numbers of crop plants of similar age and type within a defined area. If the same crop is grown on the same plot year after year, yield progressively declines. This is due in part to mineral depletion but also conditions become ideal for the crop plant's pests and parasites.

## Pest control ●●●

Pesticides are poisonous chemicals used to control organisms considered harmful to agriculture or organisms involved in disease transmission, e.g. insecticides to kill insects. Ideally a pesticide should be specific, non-persistent and should not accumulate and be passed along food chains.

Organochlorine chemicals such as DDT were persistent and remained in the environment for long periods, a property which is regarded as undesirable in a modern pesticide. The overuse of pesticides has also led to the development of resistance among many species of insects.

Despite the development of improved pesticides, such as organophosphates and synthetic pyrethroids similar to the naturally occurring pyrethrum, it is now felt that pest control is best achieved by combining various methods. These include the use of biological control agents, producing pest-resistant crops, varying cultivation techniques and where necessary the minimal, well-targeted application of highly selective pesticides. This is known as integrated pest management.

## Biological control ●●●

Biological control methods exploit natural enemies to regulate the population of pest species. A beneficial organism (the agent) is deployed against an undesirable one (the target), the aim being to bring the population of a pest down to a tolerable level. Although very effective it should be appreciated that the successful examples are limited in number, e.g. the small wasp, *Encarsia formosa*, is used to control the glasshouse whitefly, *Trialeurodes vaporarium*.

---

**Checkpoint 1**

State two advantages of large-scale monoculture.

**The jargon**

*Specificity* – a non-specific method kills beneficial as well as harmful organisms.

**Action point**

List the different types of pesticide and the types of pests they control.

**Checkpoint 2**

In terms of biodiversity what is the main disadvantage of large-scale monoculture?

**Checkpoint 3**

Explain how an insecticide which is used at a harmless concentration to wildlife can still cause the death of many animals at the top of the food chain.

**Examiner's secrets**

You should be able to give specific examples of biological and chemical control methods.

**Checkpoint 4**

Give an example of a successful glasshouse biological control.

## Deforestation

The sale of valuable timber, the freeing of land for alternative uses, clearing land for roads, etc. has meant that the trees of forests and woodland are being cut down faster than they can be replanted or regenerated naturally. The effects of deforestation are:

→ changing the composition of the atmosphere
→ soil erosion
→ destruction of natural habitats

## Fishing

Fish are an example of a renewable resource. However, if the rate at which they are removed exceeds that at which they have been produced their supply is ultimately exhausted. Fish are not generally farmed. Humans remove them from the seas with no attempt to replace stocks by breeding. Modern fishing methods have led to over-fishing. International agreement has been reached on controls such as:

→ quotas
→ mesh size of nets
→ closed seasons for fishing
→ exclusion zones

## Conservation

In recent years there has been a widespread loss of natural habitats and an increase in the number of extinct and endangered species. Conservation not only means preservation but also careful and skilful management of a complex web of biotic and abiotic resources. These include:

→ education
→ designation of national parks and nature reserves
→ planned use of land
→ legislation to protect wildlife
→ breeding in captivity
→ ecological study of threatened habitats

**Exam questions**                                              answers: page 68

1   What are the relative advantages and disadvantages of insecticides and biological control. (15 min)

2   Explain what is meant by monoculture of crops, commenting on the advantages and disadvantages of monocultures. (15 min)

**Checkpoint 5**

Which gases are most affected by deforestation?

**Checkpoint 6**

How does deforestation lead to soil erosion?

**Checkpoint 7**

Name one non-renewable energy resource.

**Test yourself**

What effects would over-fishing have on the numbers of zooplankton and phytoplankton?

**Checkpoint 8**

What is the principle behind regulating mesh size?

**Action point**

Describe the relative advantages and disadvantages of the various different management resource methods.

# Answers
## Energy and the environment

## Autotrophic and heterotrophic nutrition

### Checkpoints

1  Autotrophic: energy in; synthesis of organic compounds. Heterotrophic: energy out; breakdown of organic material.
2  Food spoilage.
3  To increase chance of finding another host.
4  Lichens, etc.
5  Holozoic.

### Exam questions

1  Apple tree (A); HIV (H, P); dog (H); nitrogen-fixing bacteria (A); *Penicillium* (H, S); tapeworm (H, P).
2  (a)  (i)   So that spores are carried a greater distance by wind currents.
         (ii)  They are aerobic and obtain oxygen on the surface.
         (iii) To increase surface area for release of enzymes.
    (b)  Digestion occurs outside the cells; digested products are absorbed into the body.

## Energy flow in the ecosystem

### Checkpoints

1  Limpets – population; temperature – abiotic.
2  Niche: position occupied and lifestyle.
   Habitat: position occupied only.
3  Passed out into space.
4  Passed on to other herbivores and decomposers.
5  The amount of energy decreases by up to 90% at each step in the food chain so more energy is available at the lower levels.
6  A protein-rich diet is more readily and efficiently digested; no energy-consuming symbiotic microbes in their digestive tract; faeces contain less undigested matter.

### Exam questions

1  (a)  $(1.8 \times 10^4$ divided by $3 \times 10^6) \times 100$ (for %) = 0.6.
   (b)  Some of the energy is of the wrong wavelength; some is reflected.
   (c)  Energy is lost at each trophic level so the biomass of animals in the higher levels is less (use the figures 100 and 1800 appropriately in your answer).
2  Your answer should include the following points.
   • The pyramid of energy is the amount of energy at each level and it is necessary to have an assessment of the amount of energy entering each level per unit time.
   • Discuss energy losses from each level.
   • Primary producer productivity can be increased by intensive planting and using fertilizers.
   • Use of monocultures simplifies the pattern of energy flow.

• Discuss effect of cropping and deforestation.
• Loss of habitat, fewer pollinators.
• Intensive stock rearing results in an increase in primary consumers.
• Use of pesticides to reduce unwanted primary consumers.
• Predator–prey relationships affected, e.g. killing of foxes may cause increase in rabbits.
• Effect of ploughing, chemicals, etc. on detritivores.

## Food chains and food webs

### Checkpoints

1  Sunlight.
2  Decomposers are saprobiontic microorganisms; detritivores are larger animals, such as earthworms, which digest food internally rather than externally.
3  Some consumers feed on more than one kind of organism and a food web is therefore a more realistic representation of feeding relationships.
4  Lost in metabolism, respiration and excretion.
5  Small mammals have a large surface area to volume ratio and as a result they lose heat rapidly. An increased metabolic rate is needed to generate heat and maintain body temperature.

### Exam questions

1  (a)  Mosquito larva, tadpole.
   (b)  Algae.
   (c)  Fish, mosquito larva, hydra, tadpole.
   (d)  Hydra, fish, tadpole.
2  (a)  (i)   Energy is reflected from the shiny cuticle of the leaf; some light passes through the leaf without being absorbed; only some of the energy is of the correct wavelength for photosynthesis.
         (ii)  It is used by the plant itself for respiration, some is lost as heat.
    (b)  Only about 10% of the energy at any trophic level is available to the next, the rest is lost in metabolism. Only a proportion of the energy is therefore passed on and there is insufficient energy available to support more than five levels.

# Ecological pyramids

## Checkpoints

1 Tertiary consumer (top), secondary consumer, primary consumer, producer (bottom).
2 Organism is heated to dryness in an oven.
3 It may not be representative of the whole.
4 The reproductive rate of microscopic organisms is much greater than that of a tree in a given period of time and therefore productivity is greater.
5 Using a calorimeter.

## Exam questions

1 (a) (i) The number of organisms in each trophic level of a food chain or web.
       (ii) The dry mass of an organism in each trophic level of a food chain or web.
       (iii) Biomass, because productivity is related to the quantity of living material rather than the number of individuals.
   (b) (i) The producer and primary consumer levels are inverted.
       (ii) The sample does not take into account the rapid reproduction (turnover) of the phytoplankton with its high rate of consumption and death. Mention the standing crop being snapshot of the situation and not a true picture of the average annual situation.
   (c) (i) Some wavelengths are unsuitable for photosynthesis; some are reflected from the leaf surface; some do not strike chlorophyll molecules.
       (ii) 8 divided by 8400 x 100 = 0.095%.

# Recycling of nutrients

## Checkpoints

1 In the past large quantities of dead organisms accumulated under anaerobic conditions and so were prevented from decay and the energy was locked within them.
2 There has been a 30% increase in carbon dioxide, which contributes as a greenhouse gas to global warming.
3 Oxidation.
4 They are anaerobic organisms and have no aerobic bacterial competitors.
5 Soil erosion and the leaching of nitrates.

## Exam questions

1 (a) R, nitrogen fixation; S, denitrification; T, nitrification.
   (b) Protein in plants (top left); nitrate (top right); nitrites (lower right).
2 (a) Clover is a leguminous plant with root nodules containing nitrogen-fixing bacteria; when the clover is ploughed in, the nitrogen content of the soil is increased.
   (b) Nitrogenous material adds excess nutrient to the water; accelerates the process of eutrophication (the effect is particularly severe if the river drains into a lake).

(c) Organic manure increases the humus content of the soil; improves the soil structure; provides slow release of nutrients.
3 (a) Your answer should include a concise description of the nitrogen cycle including putrefaction, nitrification, nitrogen fixation and denitrification. A well-annotated diagram would help the description.
   (b) • Humans fix atmospheric nitrogen artificially by chemical processes which convert it into fertilizers.
       • Large amounts of animal waste from stock rearing is used as manure.
       • Sewage disposal boosts organic nitrogen supplies.
       • The deliberate exploitation of microorganisms, e.g. during composting, silage production, etc.
       • Overuse of fertilizers leads to eutrophication in waterways.

## Examiner's secrets

Note that in questions 2 and 3 the word 'eutrophication' is used. This means that high nitrate levels in the water cause overproduction of algae, which decay when they die. The microorganisms of decay use up oxygen and the water becomes seriously depleted of oxygen so that it cannot support life.
You will need to be brief in your responses to 3(b) and you should allocate most of your time to 3(a). Be concise, but do not sacrifice detail and precision especially to the descriptions of the various stages of the nitrogen cycle.

# Population growth and control

## Checkpoints

1 The yeast is producing enzymes by protein synthesis.
2 They occupy different niches, e.g. different feeding niches means they have a different diet.
3 Interspecific competition.
4 It maintains the numbers of organisms in the food chain balanced at levels that the environment can support.

## Exam question

State of an ecosystem in equilibrium; result of a succession of communities; population numbers fluctuate about the carrying capacity; type of climax community dependent on abiotic factors; it is usually the woodland.

# Resource management and human influences

## Checkpoints

1 Ease of harvesting, easier preparation and maintenance, easier for combine harvesters to operate, greater yield so guaranteed market.
2 Reduction of gene pool; loss of hedgerows resulting in fewer natural predators.
3 Persistent insecticides build up through the trophic levels.

4 *Encarsia formosa* to control glasshouse whitefly.
5 Carbon dioxide and oxygen.
6 Lack of roots binding soil together so rainfall washes soil away.
7 Coal, oil, fossil fuels.
8 If the mesh is large enough, younger fish can escape to breed.

## Exam questions

1 Your answer should include the following points in a logical sequence.
  • Insect pests may be destroyed by chemicals (insecticides) or by parasites, predators (biological control).
  • Advantages of insecticides: very effective, instant remedy, can be applied on a large scale, no skill needed for application.
  • Disadvantages of insecticides: not specific and kill useful pollinating insects, insects can become resistant, residues may cause human illness.
  • Advantages of biological control: usually highly specific to one pest, cheap in the long term, effective long-term control if population equilibrium is established, no environmental contamination.
  • Disadvantages of biological control: a high level of skill and research is needed, release of exotic organisms with unknown ecological effects, relatively few successful examples since agents not known for most pest problems.

2 Monoculture means the growing of a single species of crop plants on a large scale. The advantages are:
  • ease of maintenance for ploughing, applying fertilizer, spraying, etc
  • ease of access for harvesting by large machinery
  • more efficient and cheaper to grow crops in a few large units than many small ones
  • large-scale production guarantees markets
The disadvantages are:
  • reduction of gene pool
  • loss of hedgerows resulting in aesthetic loss and also habitat for animals and biological control agents
  • disease or pest damage occurs on a much more devastating scale

**Examiner's secrets**

In both these questions there is a need for organization so that both sides of the argument are presented in each case. You would not be expected to cover every point mentioned here; however, avoid generalized terms and try to give concise answers describing a range of points.

# Metabolism, respiration and photosynthesis

Sunlight energy is captured in chemical form by autotrophic organisms during the process of photosynthesis. The energy-containing molecules produced are useful for storage, but they cannot be directly used by cells to provide the required energy. The transfer of energy from carbohydrate molecules, fats and proteins to molecules like adenosine triphosphate, which can be utilized by cells, occurs during respiration. ATP is a reservoir of potential chemical energy and it works in metabolism by acting as a common intermediate, linking energy-requiring and energy-yielding reactions.

## Exam themes

The universal role of ATP as the energy currency of all living organisms

The synthesis of ATP by the electron transport chain on the membranes of the mitochondria

Comparison of energy efficiency of aerobic and anaerobic respiration

The role of chlorophyll in trapping light energy

The light-dependent and light-independent stages of photosynthesis

Relate leaf structure to its role in photosynthesis

The structure of the mitochondrion and chloroplast and their role in ATP production

Respiratory quotient (RQ) and the relative energy values of respiratory substrates

## Topic checklist

| ○ AS  ● A2 | AQA/A | AQA/B | EDEXCEL | OCR | WJEC |
|---|---|---|---|---|---|
| The role of ATP in metabolism | ● | ● | ● | ● | ● |
| Aerobic respiration | ● | ● | ● | ● | ● |
| Anaerobic respiration and energy budgets | ● | ● | ● | ● | ● |
| Photosynthesis | ○● | ○● | ● | ● | ○● |
| Biochemistry of the light-dependent stage | ● | ● | ● | ● | ● |
| Biochemistry of the light-independent stage | ● | ● | ● | ● | ● |

# The role of ATP in metabolism

Adenosine triphosphate, ATP, is often described as the energy currency of cells. The energy released when respiratory substrates are broken down is used to build up molecules of ATP from adenosine diphosphate (ADP) and inorganic phosphate. When the molecules of ATP are hydrolyzed energy is released for reactions where it is needed in the cells.

## The structure of ATP ●●●

ATP is a nucleotide consisting of an organic base, adenine, a five-carbon sugar, ribose, and a sequence of three phosphate groups linked together.

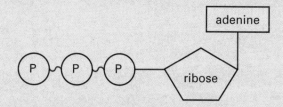

**Don't forget**

If all the energy from glucose were released at once, the temperature of the cells would increase considerably.
So when substrates are broken down in enzyme-catalyzed stages, small bursts of energy are released as ATP molecules.

## The importance of ATP ●●●

The hydrolysis of ATP to ADP is catalysed by the enzyme **ATPase** and the removal of the terminal phosphate yields 30.7 kJ mol$^{-1}$ of free energy. The addition of phosphate to ADP (phosphorylation) requires the same amount of energy. ATP is therefore a means of transferring free energy from energy-rich compounds to cellular reactions requiring it. There are two forms of phosphorylation:

→ *oxidative phosphorylation*, which occurs during cellular respiration in all aerobic cells
→ *photophosphorylation*, which occurs during photosynthesis in chlorophyll-containing cells

**Checkpoint 1**

Write the equation for the hydrolysis of ATP.

## The uses of ATP ●●●

ATP is not stored so it has to be synthesized as required. The rate of synthesis keeps pace with demand. A metabolically active cell may require 2 million ATP molecules per second. ATP provides the necessary energy for:

→ muscle contraction
→ synthesis of materials within cells
→ nerve transmission
→ active transport

**Checkpoint 2**

Give an example of a material synthesized in cells.

## The chemiosmotic theory of ATP synthesis ●●●

This theory proposes that the energy for ATP synthesis comes from the electrochemical gradient across a membrane. The same principle applies for ATP synthesis in chloroplasts and mitochondria.

**Links**

The structure of the mitochondrion is given on pages 18 and 19.

Mitochondria have an inner phospholipid membrane, which is folded to form cristae that are lined with stalked bodies. Within the inner membrane there is a mechanism which actively transports $H^+$ ions (protons) from the matrix into the space between the inner and outer membranes of the mitochondrion. This creates an electrochemical gradient of protons across the inner membrane. The process occurs as follows (see diagram below).

→ The oxidation of glucose in the matrix is coupled to the reduction of a carrier.

→ The reduced carrier passes an electron to an associated pump in the membrane, becoming reoxidized and releasing protons into the matrix.

→ The charged pump then pumps protons into the space between the inner and outer membranes.

→ Electrons with a slightly reduced energy level are transferred by a carrier to the next pump (the series of carriers and pumps is known as the electron transport system).

→ A high concentration of protons builds up in the inter-membrane space.

→ The protons are only able to flow back into the matrix via channels formed by ATPase molecules in the stalked bodies.

→ This flow acts as an energy source to combine ADP and inorganic phosphate to synthesize ATP.

→ The electrons from the last pump are taken up by oxygen, which combines with protons to form water. This is catalysed by the enzyme oxidase.

In chloroplasts the build-up of protons is inside the membrane, whereas in mitochondria the build-up of protons is outside the membrane.

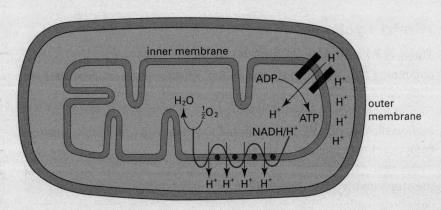

**Checkpoint 3**

What is the purpose of the cristae?

**The jargon**

The pumps in this context are large enzyme molecules that traverse the membrane.

**Checkpoint 4**

Why does ATP synthesis by this method not occur in anaerobic respiration?

**Checkpoint 5**

What name is given to the overall process of ATP production by this method?

**Action point**

Rephrase this list of points as essay-style paragraphs.

**Links**

ATP synthesis is relevant to aerobic respiration and photosynthesis, described on pages 72–73 and 76–77 respectively.

**Exam questions**                                                                    answers: page 82

1   Describe the electron transport system and explain fully how the structure of the mitochondrion:

(a) enables the electron transport chain to function

(b) ensures the synthesis of ATP

(18 min)

# Aerobic respiration

Respiration is a series of enzyme-catalysed reactions that release energy from organic molecules in order to synthesize ATP. Aerobic respiration, which involves the participation of oxygen, consists of four distinct but linked stages, glycolysis, the link reaction, Krebs cycle and the electron transport chain.

## Glycolysis

This takes place in the cytoplasm and oxygen is not required.

→ The glucose molecule is first phosphorylated to make it more reactive by the addition of two molecules of ATP to form hexose phosphate.
→ The 6-carbon hexose phosphate is split into two molecules of triose phosphate (3-carbon sugars).
→ Each 3-carbon sugar is converted to pyruvic acid.
→ Two of the steps transfer sufficient energy for the synthesis of ATP yielding a total of four ATP molecules, and since these are formed directly from phosphate compounds in the process it is known as substrate-level phosphorylation.
→ Each 3-carbon sugar initially donates two hydrogen atoms to reduce the carrier molecule NAD. If oxygen is available, $NADH/H^+$ enters the mitochondrion and brings about the synthesis of 6 ATP molecules.

## Link reaction

Pyruvic acid diffuses from the cytoplasm to the mitochondrial matrix. Pyruvic acid (3-carbon) is converted to 2-carbon acetate with a loss of hydrogen and carbon dioxide. The acetate then combines with coenzyme A to form acetyl coenzyme A (acetyl CoA).

## Krebs cycle

The acetyl CoA then enters the tricarboxylic acid, or Krebs, cycle by combining with a 4-carbon compound, oxaloacetate, to form a 6-carbon compound, citric acid; the CoA is regenerated.

The 6-carbon compound undergoes reactions during which carbon dioxide and hydrogen atoms are removed until the 4-carbon compound is regenerated via 6-carbon and 5-carbon intermediates, and is ready to combine with another molecule of acetyl CoA. Two of the steps involve decarboxylation, and four of the steps involve dehydrogenation.

For each turn of the cycle, one ATP molecule is produced by substrate-level phosphorylation and 14 are produced via the electron transport chain (oxidative phosphorylation). The hydrogen atoms produced are collected by two different carriers, with the formation of three molecules of $NADH/H^+$ (reduced NAD) and one molecule of $FADH_2$ (reduced FAD).

**Checkpoint 1**

How much energy is released from one molecule of glucose?

**Examiner's secrets**

You are not required to know the names of all the enzymes involved at each stage of glycolysis. However, you should know that dehydrogenase enzymes catalyse the removal of hydrogen atoms and decarboxylase catalyzes the removal of carbon dioxide.

**Checkpoint 2**

What is the net gain of ATP molecules in glycolysis?

**Watch out!**

For each glucose molecule broken down, the cycle turns twice. So far we have six $NADH/H^+$, two $FADH_2$ and two ATPs produced by substrate-level phosphorylation.

**Checkpoint 3**

What is the name of the enzyme which catalyzes the removal of carbon dioxide?

## Electron transport system ●●●

NAD and FAD are both **coenzymes** and are the hydrogen acceptor molecules. They pass on the hydrogen atoms through a series of carriers known as the electron transport chain and generate ATP by the proton pump mechanism.

For each pair of hydrogen atoms that pass along the chain of carriers enough energy is released for the synthesis of three molecules of ATP if NADH is involved; two if FADH.

**Links**

See also pages 70 and 71 on ATP synthesis.

**Checkpoint 4**

Explain the importance of the oxidoreductase enzymes, dehydrogenase and oxidase.

---

**Exam questions**                                            answers: page 82

1   The diagram below represents an outline of stages involved in respiration in a plant cell.

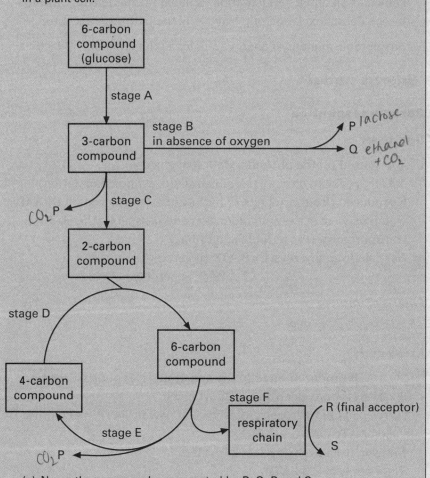

P lactose
Q ethanol + $CO_2$

**Action point**

Draw a diagram that summarizes aerobic respiration and insert at the appropriate points where, and how many, ATP molecules are formed.

(a) Name the compounds represented by P, Q, R and S.

(b) (i)  Which letter represents the stage during which most ATP is produced?

(ii) Which two letters represent separate stages where ATP is not produced?

(c) Indicate the nature of the reactions in stage F.

(10 min)

# Anaerobic respiration and energy budgets

Anaerobic respiration occurs in some microorganisms and in some tissues, e.g. muscle cells, which may be temporarily deprived of oxygen. However, without oxygen, oxidative phosphorylation will not take place and so the yield of ATP molecules will be greatly reduced. You need to be able to compare and explain the differences in yields of ATP from aerobic and anaerobic respiration.

## Anaerobic respiration

Without oxygen the carriers reduced NAD and reduced FAD cannot be reoxidized and therefore made available to pick up more hydrogen, and so the link reaction and the Krebs cycle cannot take place. Only glycolysis can occur in the absence of oxygen.

→ In vertebrate muscle cells **lactic acid** is produced. The pyruvate itself becomes the hydrogen acceptor.

→ In higher plants and yeast, **alcohol** is the product. This is also known as alcoholic fermentation. Ethanal (acetaldehyde), formed by the decarboxylation of pyruvate, is the hydrogen acceptor.

A considerable amount of energy remains in the ethanol and lactic acid.

## Energy budget

### Aerobic respiration

*Glycolysis*

2 ATP used (in the phosphorylation of glucose).
4 ATP produced directly (by substrate level phosphorylation)
6 produced (from two pairs of hydrogen atoms picked up by NAD which may enter the mitochondrion and pass along the electron transport system in aerobic conditions)
Net gain from glycolysis = **8 ATP** (if aerobic)
**2 ATP** (if anaerobic)

*Link reaction*

3 ATP (2 ×) = **6 ATP**

*Krebs cycle*

2 × 1 ATP produced directly (by substrate level phosphorylation)
2 × 11 ATP from electron transport chain
(three pairs of hydrogen atoms via NAD yields 9 ATP; one pair via FAD yields 2 ATP)
Net gain from Krebs cycle = **24 ATP**
Total net gain = **38 ATP**

### The efficiency of aerobic respiration

Each ATP molecule will yield 30.7 kJ of energy
Therefore total energy available is $38 \times 30.7 = 1\ 166.6$ kJ
The complete oxidation of glucose yields 2 880 kJ

Therefore efficiency $= \dfrac{38 \times 30.7 \times 100}{2\ 880} = 40.5\%$

**Checkpoint 1**

Under what temporary conditions do plants respire anaerobically?

**Checkpoint 2**

Why is yeast unable to respire anaerobically for an indefinite period?

**Checkpoint 3**

How can the energy locked up in lactic acid be released by muscle cells?

**Watch out!**

A total of 38 ATPs is the value most commonly quoted as being produced during aerobic respiration. In fact, 2 ATP are (usually) used up in transporting some of the metabolites across the inner mitochondrial membrane, so the value of 36 is given in some texts. As long as you know why there are these two values, either is acceptable.

**Checkpoint 4**

What is the efficiency of a car engine?

(not all the energy of the glucose molecule is captured in ATP as there is a loss of energy as heat).

## Anaerobic respiration

Energy yield = 2 ATP (glycolysis only)

Therefore total energy available = 61.4 kJ

$$\text{Efficiency} = \frac{61.4 \times 100}{2\,880} = 2.13\%$$

## Alternative respiratory substrates

In addition to glucose both fats and proteins may, in certain circumstances, be used as respiratory substrates.

### Fat respiration

Fat is a good energy store and is used as a respiratory substrate when carbohydrate is in short supply. It has to be split into its constituent molecules of **glycerol and fatty acids**, first by hydrolysis and then the glycerol is converted into a 3-carbon sugar which enters the Krebs cycle via triose phosphate, yielding ATP molecules. The long fatty acid molecules are split into 2-carbon fragments which enter the pathway as acetyl CoA. Very large numbers of ATP molecules are built up during the process, the precise number depending on the length of the hydrocarbon chain of the fatty acid, e.g. the complete oxidation of stearic acid yields nearly four times as many ATP molecules as given by the oxidation of a single glucose molecule.

### Protein

Protein is very rarely used as a respiratory substrate, usually only when all reserves of carbohydrate and fat have been used up. The protein is hydrolysed into its constituent **amino acids** and then is deaminated in the liver. The amino group is converted into urea and excreted and the residue is converted to either acetyl CoA, pyruvic acid or some other Krebs cycle intermediate and oxidized.

## Respiratory quotient (RQ)

It is possible to work out the respiratory quotient for a particular organism if the volume of carbon dioxide produced and the volume of oxygen taken up in a given period of time is known.

$$RQ = \frac{\text{volume of carbon dioxide produced}}{\text{volume of oxygen used}}$$

This information can give an indication of the type of food being used as the respiratory substrate by working out the theoretical RQ values for carbohydrates, fat and protein, and then comparing them with actual values obtained in experiments using respirometers. Carbohydrates give an RQ of 1.0, fats about 0.7 and proteins 0.9. The human RQ is usually about 0.85 indicating that a mixture of carbohydrates and fat is used for respiration.

**Action point**

Construct a simplified diagram of the respiratory pathway showing where fat and protein (as glycerol and fatty acid, and amino acids) enter the pathway.

**Checkpoint 5**

Calculate the RQ if 26 molecules of oxygen are required to completely oxidize stearic acid (a fat) and 18 molecules of carbon dioxide are evolved.

**Checkpoint 6**

Would animals living in conditions of low oxygen have a high or low RQ value?

**Action point**

Draw a diagram of a simple respirometer to show the rate of oxygen consumption in a small organism such as a locust.

---

**Exam question**                                answer: page 82

From your knowledge of the electron transport chain explain fully why glucose is not broken down completely during anaerobic respiration. (5 min)

# Photosynthesis

Photosynthesis takes place in the chloroplasts, which are found in the mesophyll cells and guard cells of green leaves. In the chloroplasts the energy of sunlight is trapped by the pigment chlorophyll. Since photosynthesis is a biochemical process involving many enzymes a suitable temperature is necessary, in addition to light, carbon dioxide and water.

## Factors affecting photosynthesis ●●●

The rate at which oxygen is given off by green plants can be used to measure the rate of photosynthesis. Blackman, in his experiments in the 1930s, showed that bubbles of oxygen were given off from the cut end of a stem of an aquatic plant such as *Elodea* when it was immersed in illuminated pond water. He showed that photosynthesis is controlled by a combination of factors and the rate is limited by whatever factor is nearest its minimum value. He called this the law of limiting factors.

### Carbon dioxide
The concentration of carbon dioxide in the atmosphere is about 0.03% and on a warm, sunny day when the light intensity and temperature are both high it can be the factor that limits the rate of photosynthesis.

### Light intensity
Increasing the light intensity causes the rate of photosynthesis to increase, up to a critical point, at which some other factor becomes limiting. Above this point, any increase in the light intensity will have no further effect on the rate.

### Compensation point
For a plant to grow it has to synthesize organic materials more rapidly than it oxidizes them in respiration. When photosynthesis and respiration occur at rates such that there is no gain or loss of organic matter, a plant is at its compensation point. At this point there is no net exchange of gases. The time taken for a plant that has been in darkness to reach the compensation point is called the compensation period.

### Temperature
Since photosynthesis is a biochemical process the enzyme-catalyzed reactions are affected by temperature.

### Water
Water is necessary for a great many metabolic reactions taking place in the plant. As a substrate in the process of photosynthesis its deficiency will reduce the rate of photosynthesis.

---

**Checkpoint 1**

Suggest how plants living in low light intensity (shade plants) benefit from having shorter compensation periods than sun plants.

**Action point**

Draw a graph showing the effect of light intensity on the rate of photosynthesis as measured by the amount of carbon dioxide exchanged over a 24-hour period.

**Action point**

Draw graphs showing the effects of light intensity, carbon dioxide concentration and temperature on the rate of photosynthesis.

## Photosynthetic pigments ●●●

The role of the photosynthetic pigments is to absorb light energy and to convert it to chemical energy. There are two types of pigments in flowering plants.

→ The **chlorophylls** absorb mainly in the red and blue-violet regions of the spectrum. There are a number of different chlorophylls with chlorophyll *a* and *b* being the most common.
→ The **carotenoids** absorb blue-violet light. They are thought to act as accessory pigments. There are two main types of carotenoids, the carotenes and the xanthophylls.

### Absorption and action spectra

Chlorophyll absorbs only certain wavelengths, reflecting others. An absorption spectrum indicates that chlorophyll absorbs wavelengths in the blue and red parts of the spectrum. However, this does not indicate whether the wavelengths are actually used. This can be shown by plotting an action spectrum of the amount of carbohydrate synthesized by plants exposed to different wavelengths of light. Since the absorption spectrum and the action spectrum for chlorophyll show a close correlation it suggests that this pigment is responsible for absorbing the light used in photosynthesis.

## Leaf structure and function ●●●

Before considering the biochemistry of photosynthesis it is necessary to consider how well the structure of the leaves of flowering plants is related to their function:

→ large surface area to capture as much sunlight as possible
→ thin, so that light can penetrate to the lower layers and there is a short diffusion path for carbon dioxide
→ the cuticle and epidermis are transparent, to allow light through
→ the palisade mesophyll cells are packed with chloroplasts and arranged with their long axes perpendicular to the surface
→ chloroplasts can move intracellularly, a process called cyclosis, allowing them to arrange themselves into the best position within the cell for the efficient absorption of light
→ chloroplasts hold the chlorophyll in an ordered arrangement
→ stomata allow carbon dioxide entry
→ intracellular spaces

---

**Checkpoint 2**

Why are leaves green?

**Checkpoint 3**

What technique is used to demonstrate the presence of the different pigments in green leaves?

**Checkpoint 4**

Why is it an advantage for a leaf to contain more than one pigment?

**Checkpoint 5**

Why do leaves appear red, orange and yellow in the autumn?

**Watch out!**

Don't confuse the absorption spectrum and action spectrum. The action spectrum shows the actual wavelengths that are used in photosynthesis.

**Action point**

Draw the action spectrum for photosynthesis and the absorption spectrum for chlorophyll on the same graph.

**Checkpoint 6**

What is the function of the leaf vein?

**Action point**

Describe in your own words how the leaf is adapted for photosynthesis by rephrasing the bullet lists as an essay-style paragraph.

---

**Exam questions**                           answers: pages 82–3

1   Describe how the structure and distribution of chloroplasts ensures the efficient trapping of light by leaves.

2   Give an illustrated account of the anatomy of a typical leaf. State the functions of the structures named.

(15 min each)

# Biochemistry of the light-dependent stage

Photosynthesis is a three-stage process. You will first need to consider how light energy is captured by the plant. The light-dependent reactions result in the formation of ATP and reduced NADP, which are then used in the light-independent stage in the reduction of carbon dioxide and the synthesis of organic compounds.

## Light harvesting

Light energy is captured by the plant using a mixture of pigments including chlorophyll. Within the thylakoid membranes of the chloroplast, chlorophyll molecules are arranged along with their accessory pigments into groups of several hundred molecules. Each group is called an antenna complex. Special proteins associated with these pigments help to funnel photons of light energy entering the chloroplast. Chlorophyll *b* is an accessory pigment molecule that passes on its energy to chlorophyll *a*, which is a primary pigment molecule and is known as the reaction centre.

There are two types of photosynthetic units, with *Photosystem I (PSI)* at a higher energy level than *Photosystem II (PSII)*.

## Light-dependent stage

During the light-dependent stage electron transport occurs as follows. The absorption of light energy by the photosynthetic pigments causes the displacement of 'excited' (high-energy) electrons which are accepted by other molecules called electron acceptors. This is an oxidation–reduction process in which the pigment molecule, the electron donor, is oxidized and the acceptor molecule is reduced. As electron transport occurs hydrogen ions are pumped from the stroma into the thylakoid discs which make up each granum. These $H^+$ flow out and down a gradient of both concentration and electrical potential difference with the production of ATP (see diagram below).

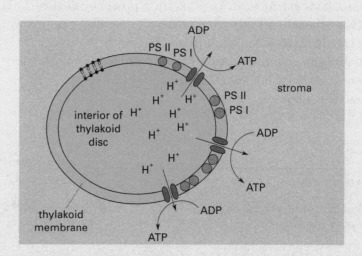

### Watch out!

Textbooks differ in their use of the term 'reduced NADP'. Some use $NADPH_2$, others $NADPH^+H^+$ (the correct terminology used by the Biochemical Society in their booklet 'Biochemistry across the school curriculum'). Throughout this section the term 'reduced NADP' will be used.

### The jargon

Radiant energy comes in discrete packets called *quanta*. A single quantum of light is called a *photon*.

### The jargon

*Photosystems* are groups of chemicals that harness the light and pass on energy.

### Checkpoint 1

What is the source of electrons for Photosystem II?

### Checkpoint 2

Which enzyme is involved in ATP synthesis?

The light-dependent stage occurs in the thylakoids of the chloroplasts and involves the splitting of water by light (photolysis). In the process ADP is converted to ATP. This is known as **photophosphorylation** (see diagram below).

Electrons from chlorophyll are passed to the light-independent reaction via reduced NADP. These electrons are replaced by others from the water molecule. Since these electrons are *not* recycled back into the chlorophyll this method of ATP production is called **non-cyclic** photophosphorylation. The process occurs as follows.

→ Light energy absorbed by PSII causes the displacement of an excited electron to a higher energy level. The electrons are received by an electron acceptor and passed via a carrier through a proton pump, from which the electrons are carried to PSI. The energy lost by the electrons is captured by converting ADP to ATP.

→ Light energy absorbed by PSI boosts the electrons to an even higher energy level and are picked up by another electron acceptor. The electrons which have been removed from the chlorophyll are replaced by the electrons from a water molecule that dissociates into protons and oxygen. The protons from the water molecule combine with the electrons from the second electron acceptor and these reduce NADP.

ATP can also be generated by **cyclic** photophosphorylation. Some electrons from PSI return to the chlorophyll directly via the electron carrier system, forming ATP in the process. These electrons are recycled, harnessing energy from light and generating ATP. No reduced NADP is produced.

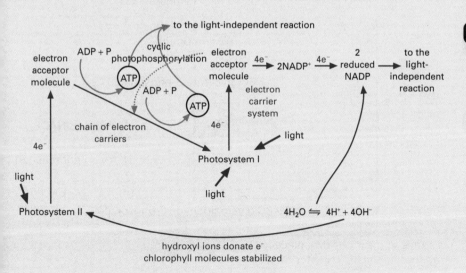

**Exam question**                                      answer: page 83

The synthesis of ATP is driven by a flow of protons. Describe where and how the required proton gradient is produced in the chloroplast. (15 min)

---

**Checkpoint 3**

With reference to the build-up of protons in the mitochondrion and chloroplast, explain how oxidative phosphorylation differs from photophosphorylation.

**Action point**

Restructure the description of non-cyclic photophosphorylation as a list of bullet points.

**Checkpoint 4**

Make a table of differences between cyclic and non-cyclic photophosphorylation.

**Checkpoint 5**

What are the three products of the light-dependent stage?

**Examiner's secrets**

In examinations this diagram may be reproduced whole or in parts. Often you are required to fill in empty boxes where key substances have been omitted.

**Watch out!**

Electrical neutrality of PSII is achieved using electrons donated from the splitting of water.

# Biochemistry of the light-independent stage

The light-independent stage occurs in the stroma of the chloroplast and involves many reactions each catalysed by a different enzyme. The reactions use the products of the light-dependent stage, ATP as a source of energy, and reduced NADP as the source of the reducing power to reduce carbon dioxide and synthesize hexose sugar.

## The Calvin cycle

The sequence of events in this stage of photosynthesis was worked out by Calvin and his associates using $^{14}C$, a radioisotope of carbon, and the unicellular algae *Chlorella*.

→ A 5-carbon acceptor molecule, ribulose bisphosphate (RuBP), combines with carbon dioxide (catalysed by Rubisco) forming an unstable 6-carbon compound.

→ The 6-carbon compound immediately splits into two molecules of a 3-carbon compound called glycerate-3-phosphate (GP).

→ GP is phosphorylated by ATP and then reduced by reduced NADP (from the light-dependent stage) to glyceraldehyde-3-phosphate (GALP).

→ Some of this 3-carbon sugar can be built up into glucose phosphate and then into starch by condensation.

→ In order that the cycle continues, the majority of the GALP formed enters a series of reactions driven by ATP which results in the regeneration of RuBP (see diagram below).

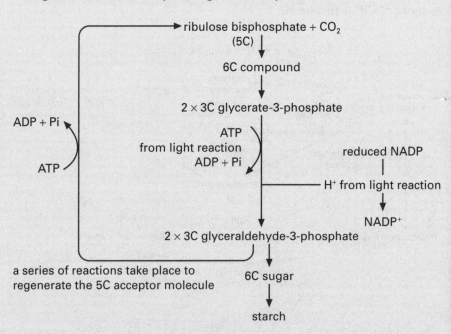

## Products of photosynthesis

The plant must synthesize all organic materials from the intermediates and products of photosynthesis. The diagram below shows an outline of the pathways involved.

### The jargon

*Rubisco* is an abbreviation for the enzyme ribulose bisphosphate carboxylase oxygenase!

### Watch out!

GP is also known as phosphoglyceric acid (PGA).

### Action point

Describe the Calvin cycle by rephrasing the bullet lists.

### Checkpoint 1

Why is it important that RuBP is regenerated?

### Checkpoint 2

How many molecules of the 3-carbon triose phosphate are needed to regenerate three molecules of RuBP?

### Watch out!

Once supplies of ATP and reduced NADP are used up then photosynthesis ends. The light-independent reaction can continue for a short time until stores of ATP, reduced NADP and GP are exhausted.

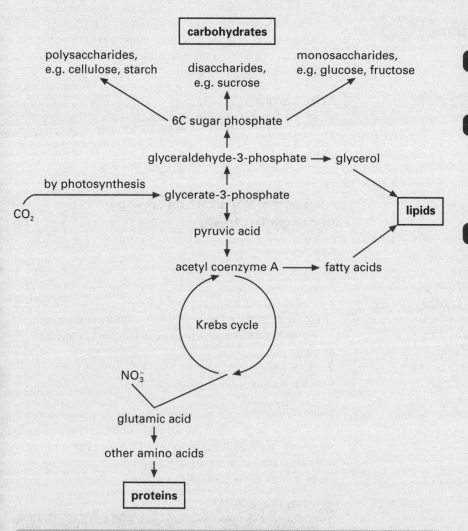

carbohydrates

polysaccharides,
e.g. cellulose, starch

disaccharides,
e.g. sucrose

monosaccharides,
e.g. glucose, fructose

6C sugar phosphate

glyceraldehyde-3-phosphate → glycerol

$CO_2$ → by photosynthesis → glycerate-3-phosphate

pyruvic acid

acetyl coenzyme A → fatty acids

lipids

Krebs cycle

$NO_3^-$

glutamic acid

other amino acids

proteins

**Checkpoint 3**

Name a polymer of glucose.

**Checkpoint 4**

What must glycerate-3-phosphate be converted to before it can enter the Krebs cycle and become part of the process of amino acid synthesis?

**Checkpoint 5**

Which one intermediate of photosynthesis and one intermediate of respiration is needed to produce lipids?

## Mineral nutrition

Mineral ions are required for the synthesis of compounds needed for the growth of the plant. Macronutrients, e.g. potassium, sodium, magnesium, calcium, nitrogen and phosphorus, are required in substantial quantities but the micronutrients, e.g. manganese and copper, are needed in much smaller amounts.

→ Nitrogen is transported as nitrates in the xylem and as amino acids in the phloem. It is needed for amino acid synthesis for proteins and nucleic acids.

→ Phosphorus is transported as inorganic phosphate ions in the xylem and as sugar phosphate in the phloem for ATP, nucleic acid and phospholipid production.

→ Magnesium ions are transported in the xylem and are used to make chlorophyll.

**Checkpoint 6**

Why do plants deficient in magnesium tend to have very yellow leaves?

**Links**

Recall the nitrogen cycle on page 60 and the uptake of minerals by roots on page 45.

**Examiner's secrets**

You only need to learn about the three minerals described in the text.

**Exam questions**                                    answers: page 84

1   Describe the sequence of reactions by which carbohydrate is produced in the light-independent stage of photosynthesis, indicating where in the cell these reactions take place. (15 min)

2   Show how the plant utilizes carbohydrates from the dark stage to make other products. Indicate the function of these products. (15 min)

# Answers
## Metabolism, respiration and photosynthesis

## The role of ATP in metabolism

### Checkpoints

1 $ATP \Rightarrow ADP + P_i + energy$.
2 Protein synthesis, DNA replication, polysaccharide from monosaccharide, etc.
3 Increased surface area for enzyme reactions.
4 No oxygen to accept hydrogen atoms so they build up.
5 Oxidative phosphorylation.

### Exam questions

1 Your answer should contain the following points in a logical sequence.
  • The energy of electrons is used to drive proton pumps.
  • The electrons are obtained from the oxidation of reduced NAD, which drives three pumps, or from the oxidation of reduced FAD, which drives two pumps.
  • Electrons are conveyed from pump to pump by carriers.
  • In order to function pumps and carriers must be assembled in proximity to one another.
  • The membrane forming the cristae perform this function.
  • The enzymes of the Krebs cycle are situated in the matrix.
  • The enzymes provide most of the reduced NAD (and FAD).
  • Oxygen is the acceptor of the final low energy electrons, with the formation of water.
  • The pumps create a proton gradient across the membrane of the cristae.
  • There is a high concentration of protons in the intermembranal space and a low concentration in the matrix.
  • ATP is formed by the enzyme ATPase.
  • ATPase is a transmembrane protein in the cristae.
  • The flow of protons down the gradient through the enzyme causes the synthesis of ATP and its release into the matrix.

## Aerobic respiration

### Checkpoints

1 2800 kJ.
2 2 (input of 2, 4 ATP produced).
3 Decarboxylase.
4 The redox reactions: when a substance is oxidized it loses electrons and when it is reduced it gains electrons; in this way electrons pass along from carrier to carrier.

Dehydrogenase catalyzes the removal of $H^+$ from a substrate and makes it available to be taken up by the electron transport system. Oxidase removes $H^+$ at the end of the electron transport chain.

### Exam questions

1 (a) P, carbon dioxide; Q, alcohol (ethanol); R, oxygen; S, water.
  (b) (i) F. (ii) A and D.
  (c) Oxidation–reduction reactions.

## Anaerobic respiration and energy budgets

### Checkpoints

1 Roots in waterlogged soil.
2 Its product, ethanol, reaches a toxic level.
3 Lactic acid is transported to the liver and converted back to pyruvate; some is oxidized to carbon dioxide and water and the rest is converted to glucose and stored as glycogen.
4 25%.
5 RQ = 0.69.
6 High RQ.

### Exam question

There is no oxygen available to remove $H^+$ ions/there is no oxygen to remove electrons. When the electrons come off the carrier system there is no final electron acceptor and so water cannot be formed.

## Photosynthesis

### Checkpoints

1 So that shade plants can make more efficient use of light at low intensities.
2 The light in the green part of the spectrum is not absorbed but is reflected.
3 Chromatography.
4 To increase the range of wavelengths from which energy may be obtained.
5 Chlorophyll is the pigment that is broken down prior to leaf fall exposing the colours of the carotenoids.
6 The xylem in the vein transports water and the phloem transports sucrose.

### Exam questions

1 Your answer should include the following points.
  • The chloroplast is surrounded by a double membrane, enclosing the stroma.
  • Within the stroma are the thylakoid membranes, which enclose the thylakoid space.

- The photosystems are held on the thylakoid membranes.
- Photons of light energy produce high energy electrons in the photosystems.
- The efficient trapping of light requires large numbers of these reaction centres.
- Thus the area of the thylakoid membranes must be large.
- Its area is increased by being stacked into grana.
- Chloroplasts occur in all the mesophyll cells of the leaf but are more abundant in the palisade cells than in the spongy mesophyll cells.
- Palisade cells are uppermost in the leaf, closest to the light.
- They are densely packed cylinders with a long axis aligned to the light direction.
- Chloroplasts can migrate up and down the cylinder to obtain optimum exposure to light.

2   Your diagram should be well annotated (see diagram below).

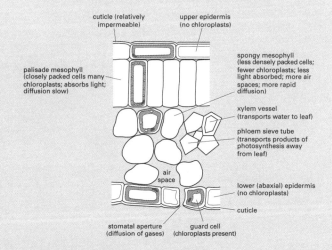

cuticle (relatively impermeable)
upper epidermis (no chloroplasts)
palisade mesophyll (closely packed cells many chloroplasts; absorbs light; diffusion slow)
spongy mesophyll (less densely packed cells; fewer chloroplasts; less light absorbed; more air spaces; more rapid diffusion)
xylem vessel (transports water to leaf)
phloem sieve tube (transports products of photosynthesis away from leaf)
air space
lower (abaxial) epidermis (no chloroplasts)
cuticle
stomatal aperture (diffusion of gases)
guard cell (chloroplasts present)

## Biochemistry of the light-dependent stage

### Checkpoints

1   OH⁻ ions from the splitting of water.
2   ATPase.

3   In the mitochondrion the build-up of protons is outside the mitochondrial membrane whereas in the chloroplast the build-up of protons is inside the thylakoid disc.
4   Cyclic: ATP produced, PSI involved only.
Non-cyclic: ATP produced, PSI and PSII involved, reduced NADP produced.
5   Reduced NADP, ATP and oxygen.

**Exam question**

1   Your answer should include 10 of the following points.
- The proton gradient is across the thylakoid membrane.
- There is a high concentration in the thylakoid space and low in the stroma.
- Protons are pumped from the stroma to the thylakoid space.
- The energy for the pumps comes from the excited electrons released from the photosystems by photons.
- The electrons are passed to and from the pump by a chain of carriers.
- Some of the electrons from PSI return as low energy electrons – the cyclic system.
- Some PSI electrons combine with protons to reduce NADP to $NADPH_2$.
- This takes place on the stroma side of the membrane and the removal of stroma protons increases the gradient.
- Lower energy electrons from PSII pass along the electron transport chain to PSI.
- This pumps more protons in the process.
- The electrons lost from PSII are replaced from the splitting of water.
- This takes place on the thylakoid space side of the membrane.
- This adds protons to the gradient.

## Biochemistry of the light-independent stage

### Checkpoints

1   So that carbon dioxide can be fixed and the cycle can continue.
2   Five.
3   Starch, cellulose.
4   Acetyl CoA.
5   Glycerate-3-phosphate and acetyl CoA.
6   Magnesium is needed to make the green pigment, chlorophyll.

## Exam questions

1 Your answer should include 10 of the following points:
- carbon dioxide is combined with a 5-carbon molecule called ribulose bisphosphate
- forming an unstable 6-carbon compound which immediately breaks down to two 3-carbon molecules called glycerate-3-phosphate (or GP/PGA)
- this is phosphorylated (with ATP)
- it then reacts with reduced NADP
- this is a reduction reaction
- glyceraldehyde-3-phosphate (GALP) is formed
- GALP reacts with ATP
- to reform ribulose bisphosphate
- or it can be converted into hexose sugar
- the reactions take place in the stroma of the chloroplast
- the stroma is the medium filling the space between the thylakoids and the outer membranes of the organelle

2 Your answer should include the following points:
- glucose is used as an energy source in cellular respiration
- it can be converted to other monosaccharides
- and combined to give disaccharides, particularly sucrose
- sucrose is the main transport molecule in phloem
- it can be synthesized into
  → polysaccharides
  → starch for energy store
  → cellulose for cell wall formation
- sugars can be converted to amino acids
- and built into plant protein
- carbohydrates can be metabolized to lipids
- use of lipids – oil storage in seeds, phospholipids in membranes, etc.

**Examiner's secrets**

Both questions 1 and 2 are straightforward recall questions. However, in question 1 a diagram alone is not sufficient. The question uses the word 'describe' and also asks you to indicate 'where' the reactions occur.
Question 2 not only asks you to give the 'functions' of carbohydrates, fats and proteins but also how the plant converts carbohydrates into the various other products.

# Regulation and control

To survive, plants and animals must react to changes in their external environment. They must have mechanisms for detecting such changes and bringing about appropriate responses. Since the structures that detect changes may be some distance from those that respond a means of communication within the body is needed. Animals have nervous systems and hormonal mechanisms, whereas plants have only the latter. Environmental stimuli must be monitored and the information fed into the appropriate system. Both plants and animals respond to the information transmitted.

## Exam themes

The principles of homeostasis in terms of receptors, effectors and negative feedback

How blood glucose concentration is regulated by negative feedback control mechanisms

Thermoregulation in mammals

The nature of communication systems in flowering plants

The commercial application of plant growth regulators

The transmission of an axon potential in a myelinated nerve

The organization of the nervous system with reference to the central and peripheral systems

The structure and functions of the mammalian brain

The structure and function of a cholinergic synapse

The role of sensory receptors in mammals in converting different forms of energy into nerve impulses

The structure of the eye and the function of its parts

The retina of the eye with reference to visual acuity, colour vision and sensitivity to different light intensities

The histology and ultrastructure of striated muscle

The sliding filament theory of muscle contraction

## Topic checklist

○ AS  ● A2

| | AQA/A | AQA/B | EDEXCEL | OCR | WJEC |
|---|---|---|---|---|---|
| Homeostasis | ● | ● | ● | ● | ● |
| The kidney | ● | ● | ● | ● | ● |
| Functioning of the kidney | ● | ● | ● | ● | ● |
| Plant growth substances | | | ● | ● | ● |
| Effect of light on plant growth | | | ● | ● | ● |
| Nerves | ● | ● | ● | ● | ● |
| Synapses | ● | ● | ● | ● | ● |
| The central nervous system | ● | ● | ● | ● | ● |
| Sensory receptors | ● | ● | | ● | ● |
| Muscles | ● | ● | ● | ● | ● |

# Homeostasis

The term 'homeostasis' is used to describe the mechanisms by which a constant internal environment is achieved. An example of the advantage of such constancy is the greater environmental freedom achieved by homoiothermic animals. The control of blood sugar is another example of homeostasis where the supply of glucose to the metabolizing cells must be kept constant even though feeding is intermittent.

## The homeostatic process

All homeostatic processes involve a **detector** (or sensor) that monitors the factor being controlled. When the detector senses a change from the normal, it informs a **controller**, a coordinating system that decides upon an appropriate method of correcting the deviation. The controller communicates with one or more **effectors**, which carry out the corrective procedures. Once the correction is made and the factor returned to normal, information is fed back to the detector which then 'switches off' (see diagram below). This is what happens in most biological control systems, i.e. the controller is no longer alerted to the deviation from the normal. This is called negative feedback.

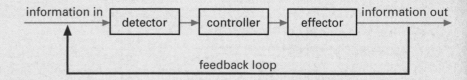

## Hormones

The endocrine system consists of a number of chemicals called hormones that regulate the body's activities.

Hormones are:

➔ produced in the tissues of endocrine glands and released into the bloodstream
➔ carried in the blood plasma to a target organ where they trigger a response. Most do this by activating enzymes

## Regulation of blood sugar

Cells are sensitive to changes in sugar level and it is therefore important to maintain 90 mg of glucose in each 100 cm$^3$ of blood. The supply of carbohydrates in mammals fluctuates because they do not eat continuously throughout the day and the quantity of carbohydrate varies from meal to meal. As cells metabolize continuously they need to be provided with a constant supply of glucose.

The liver plays a key role in glucose homeostasis. If the glucose concentration in the blood is higher than normal, the excess is stored as glycogen in the liver. If the level is less than normal, glycogen reserves are converted to glucose. Patches of pancreatic cells, called the islets of Langerhans, control this mechanism. There are two types of cells in the islets:

**The jargon**

An endocrine or ductless gland has no duct to deliver its secretions so the secretions are released into the blood as it flows through the gland.

**Checkpoint 1**

How would abnormally high concentrations of glucose in the blood affect the water content of the body?

**Watch out!**

It is not the hormones which directly interconvert glucose and glycogen. Hormones control enzymes which catalyze the conversions.

**Links**

See human insulin production by genetic engineering on page 23.

→ *α cells* produce the hormone **glucagon**, which converts glycogen to glucose

→ *β cells* produce the hormone **insulin**, which converts glucose to glycogen

Insulin also affects the rate of entry of glucose into respiring cells. It is thought to increase the rate of active absorption by triggering specific membrane receptors.

The negative feedback relationship between the hormones and glucose in blood ensures that glucose is released from the glycogen stores at a rate sufficient to match its uptake from the blood by respiring tissues.

## Temperature regulation ●●●

Homoiothermic animals have a homeostatic mechanism controlled by the hypothalamus, a small body at the base of the brain.

Within the hypothalamus is the thermoregulatory centre which has two parts, a heat gain and a heat loss centre. The hypothalamus monitors the temperature of blood passing through it and, in addition, receives nervous information from receptors in the skin about external temperature changes. Any increase in blood temperature detected by the heat loss centre causes it to trigger responses that decrease heat production and increase heat loss.

→ *Mechanisms for losing heat* in a warm environment include sweating and vasodilation (widening of arterioles in the skin).

→ *Mechanisms for retaining or gaining heat* include shivering, insulation, increased metabolic rate, vasoconstriction.

answers: page 106

**Exam questions**

1  The diagram below shows the way in which temperature is regulated in the body of a mammal.

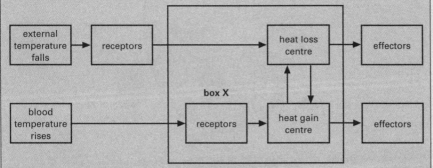

(a)  Which part of the brain is represented by box X?

(b) (i)  How does the heat loss centre control the effectors that lower the body temperature?

   (ii)  Explain how blood vessels can act as effectors and lower the body temperature.

(7 min)

2  (a)  What is meant by the term 'negative feedback'? Explain its importance in living organisms.

(b)  Explain how negative feedback can operate in mammals to regulate the blood sugar level.

(15 min)

**Checkpoint 2**

Name the temperature control centre in mammals.

**Checkpoint 3**

How does vasodilation increase heat loss?

**Checkpoint 4**

How does the heat loss centre act like a thermostat?

**Checkpoint 5**

What role do skin thermoreceptors play in temperature control?

**Test yourself**

Construct a diagram summarizing the principles of a homeostatic control mechanism and include the following terms: receptors, effectors, negative feedback, corrective mechanisms.

**Action point**

Describe the mechanisms for losing heat in a warm environment, such as a sauna.

# The kidney

The complex chemical reactions that occur in all living cells produce a range of waste products which must be eliminated from the body in a process known as excretion. In mammals the main organ of nitrogenous excretion is the kidney.

## Functions of the kidney

The kidney has two functions:

→ removal of nitrogenous metabolic waste from the body
→ osmoregulation, the mechanism by which the balance of water and dissolved solutes is regulated

## The production of urea

Urea is a poisonous chemical made by the liver. If there is too much protein in the diet any excess has to be broken down (as it cannot be stored like carbohydrates and fats). The amino acids, which make up protein, are deaminated in the liver. The reaction produces ammonia, which is quickly converted into urea. Urea is released into the blood, and travels around the body until it is removed by the kidneys.

## Structure of the kidney

The diagram below shows that within each kidney there are a number of clearly defined regions.

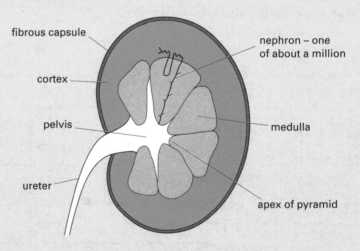

Kidneys are made up of about a million nephrons. Each nephron is made up of four functional parts: a Bowman's capsule, a proximal tubule, the loop of Henle and a distal tubule leading to a collecting duct (see diagram opposite).

Within each Bowman's capsule is a knot of blood capillaries known as the **glomerulus**. The blood supply to the nephron begins as an afferent arteriole serving the glomerulus. From the glomerulus the blood is carried by the efferent arteriole to two other capillary structures:

→ a capillary network serving the proximal and distal convoluted tubules

→ a capillary network running beside the loop of Henle and known as the **vasa recta**

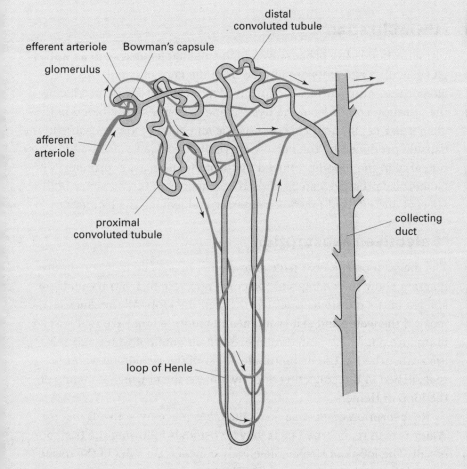

**Checkpoint 3**

Which tube carries urine to the bladder?

**Examiner's secrets**

You are expected to interpret the ultrastructure of the kidney: the arrangement of the pores in the Bowman's capsule and detail of cells from the wall of the proximal convoluted tubule.

**Check the net**

You'll find up-to-date information on the kidney at www.nephron.com/fkg.html

**Exam questions**                    answers: page 106

1   The diagram below shows part of a kidney tubule (nephron).

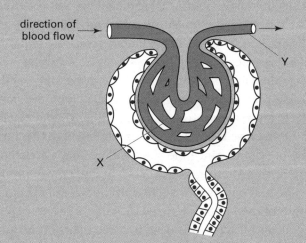

(a) What is the name given to the network of capillaries labelled X?

(b) State three consequences of constricting the diameter of blood vessel Y.

(5 min)

# Functioning of the kidney

The functioning of the nephron involves three different processes: ultrafiltration, selective reabsorption and secretion. The osmoregulatory function of the kidney is controlled by hormones.

**Checkpoint 1**

Is ultrafiltration an active or a passive process?

**Checkpoint 2**

List the components of glomerular filtrate.

**Checkpoint 3**

After filtration what are the components remaining in the blood?

**Test yourself**

Locate on the diagram of the nephron on the previous spread where:
(a) absorption of glucose and salt occur
(b) absorption of water occurs

**Checkpoint 4**

Suggest how the loop of Henle is modified in desert animals.

**Checkpoint 5**

How does the vasa recta supply essential materials to the cells of the medulla without affecting the osmotic gradient there?

**Checkpoint 6**

Where is salt concentration in the medullary tissue at its highest?

## Ultrafiltration

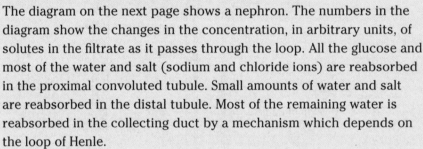

Ultrafiltration is the process by which small molecules such as water, glucose, urea and salts are filtered from the knot of capillaries, the glomerulus, into Bowman's capsule. Most of the pressure producing the filtration comes from the hydrostatic pressure of the blood in the glomerular capillaries. This pressure is amplified by the pressure in the capsule produced by the narrow efferent vessels, and also by the water potential in the blood produced by the colloidal plasma proteins. The glomerular pressure can be altered by changes in the diameter of the afferent and efferent arterioles entering and leaving the glomerulus.

## Selective reabsorption

The diagram on the next page shows a nephron. The numbers in the diagram show the changes in the concentration, in arbitrary units, of solutes in the filtrate as it passes through the loop. All the glucose and most of the water and salt (sodium and chloride ions) are reabsorbed in the proximal convoluted tubule. Small amounts of water and salt are reabsorbed in the distal tubule. Most of the remaining water is reabsorbed in the collecting duct by a mechanism which depends on the loop of Henle.

Reabsorption of glucose and salts takes place by active transport. Water is also reabsorbed passively by osmosis following the transport of salt. The loops of Henle collectively concentrate salts in the tissue fluid of the medulla of the kidney. The high concentration of salt then causes an osmotic flow of water out of the collecting ducts, thereby concentrating the urine and making it hypertonic to the blood.

The loop of Henle uses the principle of a hairpin counter-current multiplier. As fluid flows up the ascending limb, salt is actively pumped out into the tissue fluid of the surrounding medulla where a low water potential is created. The ascending limb is relatively impermeable to water while the descending limb is permeable. Water leaves the filtrate of the descending limb by osmosis and is carried away by the blood in the vasa recta. The contents of the descending limb become progressively more concentrated and reach maximum concentration at the tip of the loop; as it flows up the ascending limb the fluid becomes more and more dilute. Since the surrounding fluid also becomes more concentrated, an osmotic gradient is maintained down to the tip of the loop. The effect at one level is slight, but the overall effect is multiplied by the length of the hairpin. The result is that a region of particularly high salt concentration is produced in a deep part of the medulla. The low water potential created causes water to move from the collecting duct by osmosis.

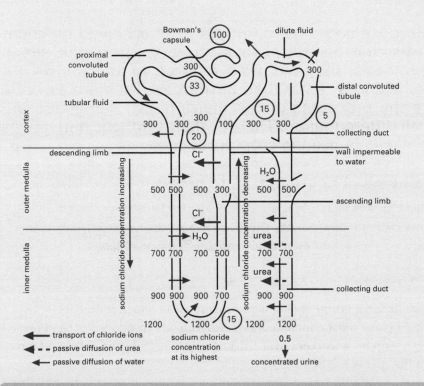

proximal convoluted tubule
Bowman's capsule
tubular fluid
cortex
descending limb
outer medulla
inner medulla
sodium chloride concentration increasing
sodium chloride concentration decreasing
dilute fluid
distal convoluted tubule
collecting duct
wall impermeable to water
ascending limb
collecting duct

100
300
33
300 300 100 300 300
20
300
5
Cl⁻
500 500 500 300
H₂O
500 500
Cl⁻
H₂O
700 700 700 500
urea
700 700
urea
900 900 900 700
900 900
1200 1200 15 1200 1200
0.5
concentrated urine

sodium chloride concentration at its highest

←  transport of chloride ions
◄ - - passive diffusion of urea
←  passive diffusion of water

## Secretion ●●●

Secretion of further substances not required by the body may take place in the distal convoluted tubule, e.g. hydrogen and hydrogencarbonate ions. This is important in the control of **plasma pH**, which must be maintained at 7.4. If the plasma pH falls, hydrogen ions are excreted by the kidney; if the plasma pH rises, hydrogencarbonate ions are secreted.

## Role of antidiuretic hormone (ADH) in osmoregulation ●●●

The amount of water reabsorbed is controlled by a feedback system. A fall in water potential of the blood may be caused by one or a combination of factors: reduced water intake, sweating, intake of large amounts of salt. The concentration of sodium chloride in the blood is an indirect indication of the volume of water in the body. The hypothalamus is sensitive to the concentration of sodium chloride in the blood flowing through it. If the water content is low, a fall in water potential is detected by **osmoreceptors** (osmotic receptors) in the hypothalamus and results in nerve impulses passing to the posterior pituitary gland which then releases **ADH**. The hormone increases the permeability of the distal convoluted tubule and the collecting duct to water and so more water is reabsorbed into the blood. Consequently the small volume of urine eventually eliminated is relatively concentrated.

### Exam questions
answers: page 106

1   (a) Explain how ultrafiltration and reabsorption remove urea from the blood without the body losing essential nutrients such as glucose.
    (b) The production of a hypertonic urine depends upon the withdrawal of water from the collecting duct into the medulla. Explain how the loop of Henle creates and maintains the conditions for this to occur.

(15 min)

**Checkpoint 7**

As fluid flows down the descending limb what happens to the salt concentration?

**Watch out!**

This is a difficult concept. In the descending limb the change in solute concentration from 300 to 1 200 is brought about as follows – the sodium and chloride ions that are in high concentration in the interstitial fluid diffuse into the descending limb. The descending limb is impermeable to the outward movement of salts and some water moves out by osmosis into the concentrated interstitial fluid. The contents of the descending limb become progressively more and more concentrated and reach a maximum concentration at the tip of the loop. In the ascending limb the change in solute concentration is brought about by the active transport of ions from the thick segment. The ascending limb is impermeable to the outward movement of water and so remains in the limb.

**Checkpoint 8**

In osmoregulation which structures act as detector, coordinator and effector?

**Checkpoint 9**

What stimulates the osmoreceptors?

**Test yourself**

Explain how the negative feedback system operates in osmoregulation.

# Plant growth substances

Because coordination in plants is achieved by chemicals that do not always move far from the site of synthesis, they are often referred to as plant growth substances rather than hormones. They are essential to the processes of cell division, cell elongation and cell differentiation. They, or their synthetic derivatives, are widely used in crop production.

## Chemical coordination in plants

Plant growth substances are chemicals that accelerate, inhibit or otherwise modify growth. There are five groups of growth substances: auxins, gibberelins, cytokinins, abscisic acid and ethene.

## Auxins

Auxins are a group of chemical substances of which **indoleacetic acid** (IAA) is the most common. Many experiments have been carried out by a variety of workers over the last century. The main points relating to IAA are as follows.

→ It is largely produced at the apices of shoots and roots.
→ Transport is polar, i.e. occurs in one direction only, away from the shoot or root tip.
→ Short-distance movement from cell to cell occurs by diffusion but long-distance transport occurs in the phloem.
→ It exerts its effect by enabling the cellulose microfibrils to slide past each other so that the cell wall can be stretched more easily as the cell takes up water osmotically during elongation.

Other effects on plant growth include:

→ promotion of root growth from cuttings
→ promotion of fruit growth, stimulates fruit development without fertilization (parthenocarpy), resulting in seedless fruit which may have a commercial advantage
→ inhibition of lateral bud development (maintains dominance of the apical bud)
→ inhibition of leaf fall

Many of these effects are commercially applied as it is possible to produce synthetic auxins. These are used:

→ to help the setting of fruit
→ as rooting hormones in powder form
→ as selective weedkillers, e.g. 2,4-dichlorophenoxyacetic acid (2,4-D)

## Gibberelins

The main points relating to gibberelins are as follows.

→ These compounds are most abundant and active in young plant organs, where they affect cell elongation.

**Examiner's secrets**

Be familiar with the experiments using oat coleoptiles (a leaf sheath).

**Checkpoint 1**

What is the advantage to a fruit tree of an early fruit drop?

**Checkpoint 2**

Explain how 2,4-D acts as a selective weedkiller (kills broad-leaved plants in cereal crops and on lawns).

**Checkpoint 3**

How may a herbicide of broad-leaved plants improve the productivity of cereal crops?

→ If applied to dwarf varieties of plants, they stimulate the rapid growth of the stem and the plant develops to normal size by increasing internode growth.

→ They are involved in the breaking of dormancy in buds and seeds. Studies with cereal grains have shown that the release of enzymes takes place from the aleurone layer surrounding the endosperm. After the absorption of water by the grain there is a marked rise in embryonic production of gibberelins, which diffuse into the endosperm, where food reserves are hydrolyzed, so providing the embryo with materials for growth. Gibberelins and auxins supplement each other in stem elongation.

Gibberelins are used commercially to promote the setting of fruit and in parthenocarpy.

## Cytokinins

The cytokinins form a group of plant growth substances that are involved in cell division..Together with auxin, they promote cell division in the apical and lateral meristems. They have also been shown to cause delay in the ageing of leaves, and are involved in the breaking of dormancy in buds and leaves.

## Abscisic acid

Abscisic acid is a growth inhibitor and appears to be antagonistic to the action of the other growth substances. Its main effect is on abscission (leaf and fruit fall). The process results from a balance between the production of auxin and abscisic acid. As the fruit ripens the level of auxin (which inhibits abscission) falls, while that of abscisic acid (which promotes abscission) increases. This leads to the formation of an abscission layer, which causes the fruit to fall.

## Ethene

Ethene is a gas that is produced by most plant organs. It appears to stimulate the rapid increase in respiration rate that occurs before the ripening of fruit. Knowledge of its effects has been used in the commercial control of ripening. Fruit is stored in oxygen-free conditions to prevent ripening, which can be induced by applying oxygen and ethene when required.

**The jargon**

An *internode* is the part of the shoot between one bud and the next.

**Links**

See germination on page 132.

**Checkpoint 4**

Why are some synthetic growth regulators more useful to us than naturally occurring substances?

**Checkpoint 5**

If you removed the apical bud from a shoot and applied cytokinin to the cut surface what would you expect to happen a few days later?

**Watch out!**

Be aware that growth substances interact with one another either by supplementing each other (synergism) or by having an antagonistic effect.

**Checkpoint 6**

Why does a ripe banana, placed near unripe fruit but not in contact with them, speed up their ripening?

**Exam question**                                          answer: page 107

Explain why auxins, cytokinins and gibberelins can be important to gardeners and farmers. (15 min)

# Effect of light on plant growth

Light affects many aspects of plant growth, from the synthesis of chlorophyll to photosynthesis and phototropic movements. A number of plant growth responses are influenced differently by light of different wavelengths. For light to have an effect it must be absorbed by a photoreceptor substance, phytochrome. Flowering is regulated by day-length.

## Phototropism

A **tropism** is a growth movement of part of a plant in response to a directional stimulus. The response to light is termed phototropism. The direction of the response is described as positive if movement is towards the stimulus, or negative if it is away from it.

Early experiments on auxins used oat coleoptiles.

→ If oat coleoptiles were subjected to unilateral illumination, they showed a positive phototropic response and grew towards the direction of the light, the bending occurring just behind the tip.
→ If the tip of the coleoptile was removed, then no response was shown to the direction of the light, and similarly no response was shown if the tips were covered.
→ If the coleoptile was covered except for the tip, then a positive phototropic response was again shown.

As a result of these experiments, it could be concluded that the tip detected the stimulus of light and that the response was made in the region just behind the tip (see diagram below).

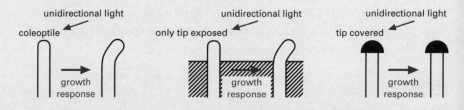

## Phytochrome

The photoreceptor responsible for absorbing light has been identified as **phytochrome**, a blue-green pigment found in very minute quantities in plants. Phytochrome exists in two forms that are interconvertible:

→ one form absorbs red light, with an absorption peak at 660 nm, and is referred to as Pr
→ the second form absorbs far-red light, with an absorption peak at 730 nm, and is referred to as Pfr

When Pr absorbs red light it is converted to Pfr. There is more red light than far-red light in natural sunlight, so Pfr tends to accumulate during the day and become slowly converted back to Pr at night. It has been shown that the phytochrome system is important in many aspects of plant growth, such as germination in some seeds, stem elongation, leaf expansion, growth of lateral roots and flowering.

---

**Checkpoint 1**

In experiments on phototropism the coleoptiles are grown in the dark except for brief periods of exposure to unilateral light. Why is it necessary to keep coleoptiles in the dark?

**Checkpoint 2**

Suggest how the response of a shoot to light is important for successful growth of a seedling.

**Checkpoint 3**

If a shoot tip is covered with tinfoil and light is shone unilaterally what will be the response?

**Checkpoint 4**

Coleoptiles subjected to unilateral light have their tips removed and placed on agar blocks for several hours. When the blocks are placed on the tops of decapitated coleoptiles but on the right side only, in which direction will the coleoptiles bend?

**Checkpoint 5**

Why does Pfr tend to accumulate during the day?

## Photoperiodism

Photoperiodism is the term used to describe the influence of the relative lengths of periods of light and darkness on the activities of plants, and day-length has been shown to have an important effect on flowering. Flowering plants can be divided into three groups according to their photoperiodic requirements prior to the production of flowers.

→ *Day-neutral plants*: flowering does not seem to be affected by day-length, e.g. tomato, cucumber, antirrhinum.
→ *Long-day plants*: flowering is induced by exposure to dark periods shorter than a critical length, e.g. petunia, spinach.
→ *Short-day plants*: flowering is induced by exposure to dark periods longer than a critical length, e.g. chrysanthemum, tobacco, poinsettia.

Flowering in short-day plants is inhibited by exposure to red light, while exposure to far-red light will bring about flowering. It seems that these plants will flower only if the level of Pfr is low enough. The situation in long-day plants is reversed and flowering is triggered by high levels of Pfr. The length of the photoperiod is less critical than the length of the dark period, and if the photoperiod is interrupted with a short period of darkness flowering still follows (see diagram below). If the dark period is interrupted by as little as 1-min exposure to light, flowering is prevented. Red light is most effective in this respect yet the effect of red light treatment can be overcome if the plant is immediately exposed to infrared light.

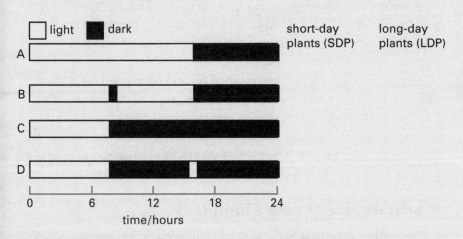

The photoperiodic stimulus is perceived by the leaves and this can be shown easily by leaving one leaf of a plant exposed while the rest of the plant is covered up. The stimulus must be transmitted in some way to the buds, which then develop flowers. The suggestion is that phyto-chrome causes a hormone to be produced which affects the buds.

**Exam questions**                                    answers: page 107

1   What is photoperiodism?

2   How does the photoperiod affect flowering?

                                                      (15 min)

---

**Checkpoint 6**

Why is it now considered unfortunate that historically plants are categorized as short-day or long-day?

**Checkpoint 7**

Define photoperiod.

**Checkpoint 8**

Using the diagram opposite state in each case whether flowering takes place in short-day or long-day plants.

**Checkpoint 9**

Which part of the plant contains phytochrome and so detects the light?

# Nerves

Unlike plants, coordination in animals is brought about by means of two separate but interconnected systems, the nervous system and the endocrine system. The former gives a rapid and short-term response while the latter gives a slower but more lasting one. The cells of the nervous system transmit nerve impulses and are called neurones. You will need to consider their structure and how they are able to transmit information at a remarkable speed in mammals.

## Structure of neurones

In the nervous system there are three types of **neurones**.

→ *Sensory or afferent*: bring impulses from the sense organs or receptors into the central nervous system.
→ *Motor or efferent*: take impulses from the central nervous system to the effector organs (muscles or glands).
→ *Intermediate or relay neurones*: receive impulses from sensory neurones or other intermediate neurones and relay them to motor neurones or other intermediate neurones.

Each cell consists of a cell body containing a nucleus and granular cytoplasm containing many ribosomes. These ribosomes are grouped together forming Nissl granules, which are concerned with the formation of neurotransmitter substances. From each nerve cell body arise fine branching structures, **dendrites**. These receive impulses from other nerve cells. Some neurones also have a long membrane-covered cytoplasmic extension, the **axon**, which transmits impulses from the cell body. At its end, an axon divides into branches which form synapses with other neurones.

Peripheral neurones are surrounded by and supported by Schwann cells. In some cases, these grow around the axons of the nerve cells to form a multilayered fatty myelin sheath, found only in vertebrate nervous systems. This acts as an electrical insulator and speeds up the transmission of impulses. The myelin sheath has thin areas at intervals, nodes of Ranvier, which are important in impulse transmission.

## Transmission of nerve impulses

→ The **resting potential** is the potential difference between the inside and the outside of a membrane when a nerve impulse is not being conducted.
→ Resting potentials are typically minus values, the minus indicating the inside is negative with respect to the outside. The membrane is said to be polarized.
→ The sodium–potassium exchange pumps (these are transmembrane proteins) maintain the concentration and an uneven distribution of sodium ions and potassium ions across the membrane.
→ The composition of the cytoplasm of an axon is very different from that of the surrounding fluid. Potassium ($K^+$) and organic anions ($COO^-$) are higher inside the neurone while the concentration of

**Links**

Because of their close association with particular organ systems, the activities of certain endocrine glands are dealt with earlier in this section and on pages 136–7.

**Checkpoint 1**

In table form list the differences between hormonal and nervous communication using the following headings: origin of stimulus, nature of stimulus, means of transmission, destination of stimulus, receptor, speed of transmission, effects, duration.

**Checkpoint 2**

Describe the direction of the impulse through a neurone.

**Checkpoint 3**

Why do myelinated axons conduct impulses faster than non-myelinated fibres of the same diameter?

**Checkpoint 4**

Is the net charge inside the membrane of a neurone positive or negative?

**Checkpoint 5**

List in the correct order the sequence of changes in the membrane of an axon between the resting potential condition, the passage of an impulse and the re-establishment of a resting potential.

sodium (Na$^+$) is higher outside. The membrane is more permeable to K$^+$ than any of the others.

→ Since the concentration of K$^+$ ions is higher than outside, they diffuse out. This outward movement of positive ions means that the inside becomes slightly negative.

→ Nerve impulses are due to changes in the permeability of nerve cell membrane to K$^+$ and Na$^+$ that leads to changes in the potential difference across the membrane and the formation of action potential.

→ Suitable stimulation of an axon results in change of potential across the membrane from a negative inside value of about −70 mV to a positive inside value of +40 mV. This change is called an **action potential**. The membrane is said to be **depolarized**.

→ The action potential is the result of a sudden increase in the permeability of the membrane to Na$^+$. This allows a sudden influx of Na$^+$ which depolarizes the membrane.

→ A fraction of a second after this depolarization the K$^+$ ions diffuse out and repolarize the membrane. There is an overshoot of K$^+$ leaving as the K$^+$/Na$^+$ pump restores the ionic balance. This is called the **refractory period** during which another action potential cannot be generated, so ensuring a unidirectional impulse and limiting frequency.

The process may be summarized as follows. The action potential causes a small electric current across the membrane and as a portion of the membrane is depolarized, depolarization of the next portion is initiated. There is a series of local currents propagated along the axon. The sodium pump is active all the time and behind the transmission; this pump restores the resting potential. Once the resting potential is restored, another impulse can be transmitted. As the impulse progresses the outflux of K$^+$ causes the neurone to be repolarized behind the impulse. There is an overshoot of K$^+$ leaving as the K$^+$/Na$^+$ pump restores the ionic balance. During the period between depolarization and repolarization impulses cannot pass. This is known as the refractory period.

When axons have a myelin sheath there is no contact between the axon membrane and the tissue fluid except at the nodes of Ranvier. The result is that the impulse jumps from one node to the next, speeding the overall passage along the axon. Also, the larger the diameter of the axon, the greater the velocity of transmission (see diagram below).

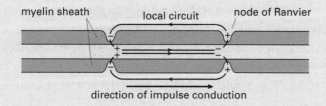

myelin sheath    local circuit    node of Ranvier

direction of impulse conduction

**Action point**

Draw a labelled diagram of a generalized neurone and show the direction of the nerve impulse.

**Checkpoint 6**

What is the source of energy used to establish a resting potential?

**Checkpoint 7**

Name two features of mammalian neurones that increase the speed of transmission of impulses.

**Checkpoint 8**

What is the importance of the refractory period?

**Checkpoint 9**

An action potential is an 'all or nothing' response. What does this mean?

**Exam question**                                    answer: page 108

Give an account of the general structure of the mammalian motor neurone and describe the way in which the impulse is transmitted along the neurone. (17 min)

# Synapses

Neurones are not in direct contact with each other but are separated by tiny gaps known as synapses. The main role of the synapse is to convey action potentials between neurones. A range of drugs function by interfering with the neurotransmitters involved in synaptic transmission.

## Structure of a synapse

Most junctions between neurones take the form of chemical synapses. Branches of axons lie close to dendrites of other neurones but do not touch; there is a gap of about 20 μm between them. When impulses are transmitted, this gap is crossed by the secretion of a neurotransmitter from the axon membrane (presynaptic membrane), which diffuses across the space to stimulate the dendritic membrane (postsynaptic membrane). The structure of a synapse is shown below.

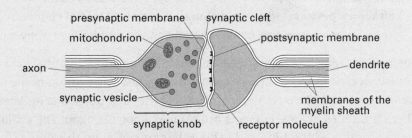

direction of transmission
of action potential

presynaptic membrane     synaptic cleft
mitochondrion                               postsynaptic membrane
axon                                                            dendrite
synaptic vesicle
                                                    membranes of the
                                                    myelin sheath
synaptic knob              receptor molecule

## Synaptic transmission

The arrival of the impulses at the synaptic knob alters its permeability allowing calcium ions to enter. The influx of calcium ions causes the synaptic vesicles to fuse with the presynaptic membrane so releasing the transmitter (acetylcholine) into the synaptic cleft. When the transmitter diffuses across the gap (synaptic cleft) it attaches to a receptor site on the postsynaptic membrane, depolarizing it and so initiating an impulse in the next neurone.

The postsynaptic membrane contains specific protein receptors with which the transmitter molecules combine. Once combined, protein channels open up in the membrane, allowing sodium ions to diffuse from the cleft into the postsynaptic neurone. Acetylcholine, when released, is quickly destroyed by enzymes in the synaptic cleft, so its effect is limited and the merging of impulses is prevented. If insufficient acetylcholine is released, the postsynaptic membrane will not be stimulated. The enzyme that destroys acetylcholine is called cholinesterase. The resulting choline and ethanoic acid diffuse back across the synaptic cleft for further use.

## Functions of synapses

→ Transmit information between neurones.
→ Pass impulses in one direction only.
→ Act as junctions.
→ Filter out low-level stimuli.

### Checkpoint 1

What general name is given to chemicals which transmit messages across a synapse?

### Checkpoint 2

How can there be an unlimited supply of acetylcholine?

### Checkpoint 3

By what process do ions cross the post-synaptic membrane when a neurotransmitter substance has caused ion-specific channels to be opened?

### Checkpoint 4

What is the function of organelles such as mitochondria and the Golgi apparatus in the synaptic knob of neurones.

## Effects of drugs ●●●

Drugs which interfere with synaptic transmission can be classified into two types:

→ excitatory drugs amplify the process of synaptic transmission
→ inhibitory drugs inhibit synaptic transmission

Nicotine mimics natural transmitters and binds to their specific receptors, blocking their action. Caffeine raises cell metabolism leading to the release of more neurotransmitters. Organophosphorous insecticides inhibit the cholinesterase enzyme.

**Checkpoint 5**

Describe the effect of an insecticide on synaptic transmission and the symptoms exhibited by the affected insect.

**Checkpoint 6**

Is caffeine an excitatory or inhibitory drug?

---

**Exam questions**                                         answers: page 108

1  The diagram below represents a synapse in the nervous system of a mammal.

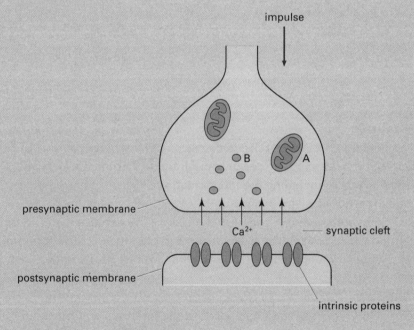

(a) (i)   Name structures A and B.
    (ii)  Name the contents of B.
(b) The arrival of a nerve impulse changes the permeability of the presynaptic membrane, allowing calcium ions to rapidly diffuse in as shown by the arrows on the diagram. Describe the effect caused by this influx of ions.
(c) Sodium ions also play a vital part in this process.
    (i)   Use the diagram to name the parts of the synapse which can act as a channel for sodium ions.
    (ii)  Suggest how these channels are opened.
    (iii) Describe the effect their opening has on the postsynaptic membrane.
(d) Explain fully why structure A is found abundantly in the presynaptic region.

                                                              (14 min)

# The central nervous system

The mammalian nervous system is dual in nature. The central nervous system (CNS) consists of the brain and spinal cord, which coordinate and control the activities of the animal. The peripheral nervous system, the nerves and ganglia, forms the connecting link between the organs and the CNS. Many body functions and actions are controlled by reflex actions, which involve both parts of the nervous system.

## Structure and function of the brain

The brain is divided into three main regions: forebrain, midbrain and hindbrain (see diagram below).

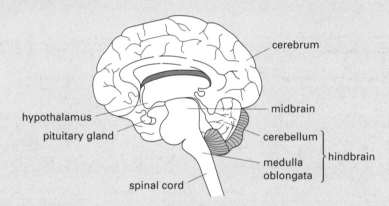

### Forebrain

The dorsal part of the forebrain consists of two large cerebral hemispheres forming the **cerebrum**. All the higher mental activities take place here including the interpretation of sensory information, thought processes and memory.

The **hypothalamus** is found at the base of the cerebrum and contains centres which control the basic drives and emotions. It also monitors the composition of the blood and the temperature of the body.

### Midbrain

This acts as an important link between the forebrain and the hindbrain.

### Hindbrain

This consists of two main parts.

→ The **medulla oblongata** contains centres controlling reflex activities such as ventilation rate, heart rate and blood pressure.
→ The **cerebellum** is concerned with the coordination of body movements and postural responses. Impulses from the motor centres in the forebrain pass to the spinal cord via the cerebellum. It also receives sensory impulses and its function is to moderate the motor impulses to produce appropriate muscular movements.

**Watch out!**

Be able to describe the correct function of each part of each region of the brain.

**Checkpoint 1**

Which part of the brain is concerned only with involuntary actions?

**Links**

See osmoregulation on page 91 and thermoregulation on page 87.

**Checkpoint 2**

Which is the largest part of the human brain?

## Spinal cord ●●●

This is the portion of the CNS which passes through the vertebral column and from which most of the **peripheral nerves** originate. It consists of a central area of grey matter, consisting mainly of nerve cell bodies, and a surrounding area of white matter, which consists of nerve fibres. Sensory fibres of the peripheral system enter the cord via the dorsal routes. The cell bodies of the sensory fibres are found in the dorsal root ganglia, which lie alongside the spinal cord. The motor fibres leave via the ventral roots (see diagram below).

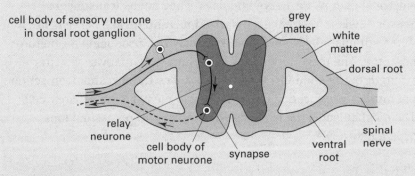

cell body of sensory neurone in dorsal root ganglion · grey matter · white matter · dorsal root · relay neurone · cell body of motor neurone · synapse · ventral root · spinal nerve

The functions of the spinal cord are to relay impulses in and out of any particular point along the cord, and to relay impulses up and down the body, including to and from the brain.

## Reflex arc ●●●

A reflex arc is a rapid, automatic response resulting from nervous impulses initiated by a stimulus. They are generally protective in function.

The following sequence describes a typical reflex action, flexion of the arm in response to a hot surface:

→ stimulus – the hot surface
→ receptor – temperature and pain receptors in the skin
→ sensory neurone transmits impulse to spinal cord
→ relay neurone connects sensory neurone to motor neurone
→ motor neurone transmits impulse to effector (muscle)
→ response – arm muscles contract and hand is removed from surface

## Autonomic nervous system ●●●

The autonomic nervous system is divided into two parts: the **sympathetic** system, which generally stimulates effectors, and the **parasympathetic** system, which generally inhibits effectors. They work antagonistically to each other, for example the sympathetic system increases ventilation rate, while the parasympathetic system decreases ventilation rate.

**Checkpoint 3**

Why is the grey matter darker in colour than the white matter?

**Test yourself**

Add to the diagram opposite a drawing of the arm. Extend the appropriate neurones to the hand and biceps muscle. Number the six points listed in the text below the diagram and insert them on your version.

**Checkpoint 4**

Define a reflex.

**Checkpoint 5**

Define a reflex arc.

**Checkpoint 6**

Some reflexes, e.g. knee jerk reflex, do not involve a relay neurone and so the brain is not 'informed'. Why is this an advantage?

**Action point**

Make notes on the differences between the sympathetic and parasympathetic systems.

**Exam question**                                    answer: page 109

Outline the arrangement of the main regions in the mammalian brain, briefly indicating the functions of each. (15 min)

# Sensory receptors

For the nervous system to carry out its function effectively it is dependent upon a continuous input of information from inside the body and from the environment. This input is initiated by sensory receptors, which range from specialized sensory cells to the most complex sense organs such as the mammalian eye.

## Mode of action of receptors

Sensory receptors detect one form of energy and convert it into electrical energy, i.e. nerve impulses. They act as **transducers**. Sensory cells at rest have a polarized membrane like nerve cells. When stimulated, the membrane depolarizes, producing a generator potential. The magnitude of the generator potential varies with the strength of the stimulus. When this reaches the threshold, an action potential is produced in the sensory axon leaving the sensory cell. The mechanism is the exchange of sodium and potassium ions, as in a neurone.

## The eye

The structure of the eye and the function of its parts is shown below.

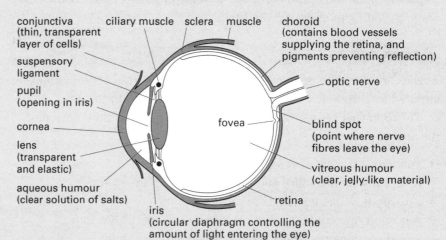

conjunctiva (thin, transparent layer of cells)   ciliary muscle   sclera   muscle   choroid (contains blood vessels supplying the retina, and pigments preventing reflection)

suspensory ligament

pupil (opening in iris)

cornea

lens (transparent and elastic)

aqueous humour (clear solution of salts)

optic nerve

fovea

blind spot (point where nerve fibres leave the eye)

vitreous humour (clear, jelly-like material)

retina

iris (circular diaphragm controlling the amount of light entering the eye)

### Control of the amount of light entering the eye
The iris diaphragm controls the amount of light entering the eye. In bright light:

→ more photoreceptor cells in the retina are stimulated
→ more impulses pass along sensory neurones to the brain
→ the brain sends impulses to the iris diaphragm
→ circular muscles contract and radial muscles relax
→ the pupil constricts and less light enters the eye

### Focusing
The condition of the eye when focused on a near object is shown in the diagram at the top of the next page.

**Checkpoint 1**

Which two parts of the eye are involved in focusing?

**Checkpoint 2**

Which part of the eye contains the sensory receptors?

**Checkpoint 3**

What is the function of the choroid layer?

**Test yourself**

Describe the mechanism of control in dim light.

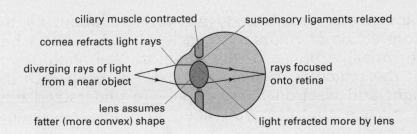

ciliary muscle contracted — suspensory ligaments relaxed

cornea refracts light rays

diverging rays of light from a near object → rays focused onto retina

lens assumes fatter (more convex) shape — light refracted more by lens

## The retina

The retina possesses the photoreceptor cells. These are of two types, **rods** and **cones**, and their structure is shown in the diagram below.

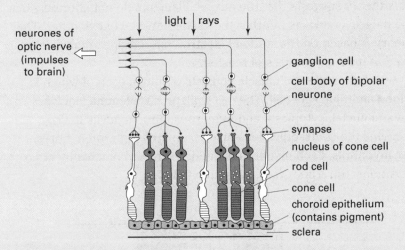

light rays

neurones of optic nerve (impulses to brain)

ganglion cell

cell body of bipolar neurone

synapse

nucleus of cone cell

rod cell

cone cell

choroid epithelium (contains pigment)

sclera

Rod cells are distributed throughout the retina whereas cone cells are concentrated in and around the fovea. Many rod cells share a single neurone and therefore give poor visual acuity (sharpness of vision). Each retinal photoreceptor cell contains a photosensitive pigment. In a rod cell this is rhodopsin or visual purple. Light absorbed by the rod cell causes rhodopsin to split into its constituent parts, opsin and retinal, in a process called bleaching. It is this chemical change that initiates an action potential along the neurone joining the rod cell to the brain.

## Colour vision

It is only cone cells that are sensitive to different colours. It is thought that they are of three types, any one being sensitive to blue, green or red light. It is the degree of stimulation of each of these three types which provides the full range of colours visible to organisms with colour vision. This theory is called the trichromatic theory of colour vision.

**Exam question**                                    answer: page 109

From your knowledge of the retina explain why two small objects close together can be more easily distinguished by rods than cones. (3 min)

**Test yourself**

Describe the condition of the eye when focused on a distant object.

**Checkpoint 4**

Why do some people hold a book at arm's length in order to read? Why does a bright light help?

**Checkpoint 5**

Make up a table to compare rods and cones using the headings: relative numbers, distribution in retina, visual acuity, light sensitivity, ability to discriminate light of different wavelengths.

**Checkpoint 6**

Which part of the retina contains only cone cells?

**Checkpoint 7**

The diagram of the rod cells shows there are numerous mitochondria present. Suggest how the resynthesis of rhodopsin occurs.

**Checkpoint 8**

Suggest how we can see other colours besides blue, green or red.

# Muscles

Movements of the body are brought about by the contraction of voluntary muscles, which are attached to movable bones of the skeleton. You should look at the structure of skeletal muscle as seen with the light and electron microscopes. To understand how a muscle contracts you need to study the sliding-filament theory.

## The sliding-filament theory of muscular contraction

This theory suggests that the muscle filaments do not shorten when the muscle contracts but that they slide between one another. The theory is based on the study of electron micrographs of muscles fixed at different degrees of tension.

Striated or skeletal muscle consists of fibres, each of which is a long multinucleate cell. The cell wall, or **sarcolemma**, encloses several nuclei, cytoplasm and numerous mitochondria. It also contains the contractile elements, which are long protein fibres, the myofibrils. Each myofibril is divided into sarcomeres by cross partitions, the Z lines (see diagram below).

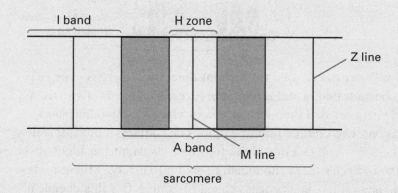

In each sarcomere there are two types of protein fibre:

→ thin **actin** strands project into the sarcomere from the Z lines
→ thicker **myosin** strands between them, which are fixed in position by molecular bonds (see diagram below).

**The jargon**

A *myofibril* consists of a number of repeated sarcomeres.

**Checkpoint 1**

Is the A band made up of both actin and myosin or only one of them?

**Checkpoint 2**

What makes up the I band?

**Checkpoint 3**

Why does a muscle fibre contain light and dark bands?

**Action point**

By means of a bullet list describe the changes that take place when a muscle contracts.

**Check the net**

You'll find information about the role of calcium ions in muscle contraction at shef.ac.uk/uni/projects/mc/excont.html

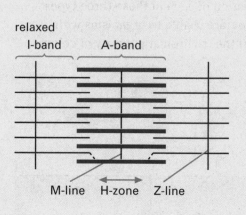

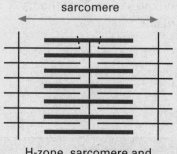

H-zone, sarcomere and I-band all shorten the A-band, does not change in length

During contraction the actin and myosin filaments slide together to form a shorter sarcomere (see diagram on previous page). The filaments slide together because of the formation of cross bridges that alternatively build and break during contraction.

## The ratchet mechanism

The 'ratchet mechanism' may be used to explain how the actin and myosin filaments slide past one another.

→ There are cross bridges between the actin and myosin filaments.
→ When an impulse reaches the neuromuscular junction, acetylcholine is released, which causes the sarcolemma of the muscle to be depolarized. This causes the release of calcium ions, which activate the protein **troponin**. This displaces another protein, **tropomyosin** (which has been blocking the actin filament), away from attachment sites on the actin molecule so myosin and actin can be linked by cross bridges (see diagram below).
→ The bulbous heads along the myosin filaments form these bridges.
→ Each myosin filament has a number of bulbous heads.
→ The clubbed heads of the myosin molecule swing back and forth with the breaking and reforming of bonds.
→ The bulbous heads progressively moves the actin filament along, as it becomes attached and detached.
→ The process receives energy by the hydrolysis of ATP.
→ Phosphocreatine is used to regenerate ATP.

**The jargon**

The *neuromuscular junction* is the point where the effector nerve meets a skeletal muscle.

**Action point**

Describe the changes that occur in the appearance of a myofibril on contraction.

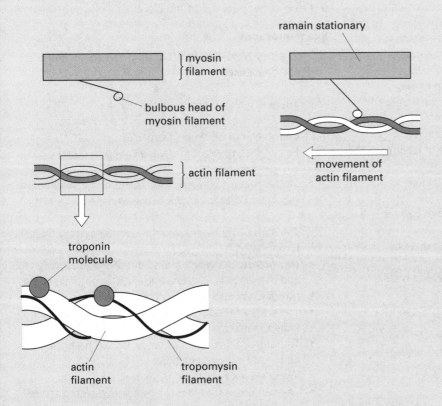

troponin molecule

actin filament

tropomysin filament

**Checkpoint 4**

Using the diagram opposite how would you describe the proteins tropomyosin and troponin, fibrous or globular?

**Examiner's secrets**

The sequence of events for the stimulation of a muscle is much the same as the mechanism of synaptic transmission.

**Exam question**                                   answer: page 109

Give an illustrated account of the sliding-filament theory of muscle contraction. (15 min)

# Answers
## Regulation and control

## Homeostasis

### Checkpoints

1 Dehydration.
2 Hypothalamus.
3 Superficial arterioles dilate allowing more blood to run through the surface capillaries increasing heat loss through radiation.
4 It switches on the heat loss mechanisms when the temperature of the blood is higher than normal. The heat gain centre is inhibited by the heat loss centre but becomes active when receptors in the skin signal that the environment is getting cooler. Heat production mechanisms and conservation mechanisms are also switched on.
5 They monitor external temperature changes.

### Exam questions

1 (a) Hypothalamus.
   (b) (i) Nerve impulses.
       (ii) Vasodilation of arterioles brings more blood through the surface capillaries, with the loss of heat by radiation.
2 Your essay should cover 10 of the following points in a logical sequence.
   (a) Negative feedback means that an increase in some factor causes changes to occur which operate to bring about a decrease in the original factor. This provides a self-regulatory mechanism for maintaining a constant level in the factor. This is the basis of homeostasis and it is important in living organisms that all internal processes fluctuate only marginally from the norm.
   (b) The factor is the level of glucose in the blood.
   • Any fluctuation in glucose concentration initially affects the level of insulin secretion by the β cells of the islets of Langerhans in the pancreas.
   • Increased glucose increases insulin secretion.
   • Increase in insulin increases the uptake of glucose by muscle cells.
   • This promotes oxidation of glucose in the liver or muscle.
   • This promotes the conversion to glycogen by liver cells and inhibits glycogen breakdown.
   • Decrease in insulin reduces uptake of glucose by cells.
   • Lowered blood glucose levels stimulate islet α cells to release glucagon.
   • Glucagon stimulates the breakdown of glycogen.

## The kidney

### Checkpoints

1 Cortex.
2 Medulla.
3 Ureter.

### Exam questions

1 (a) Glomerulus.
   (b) Pressure at X would increase.
       Rate of flow of blood decreases.
       Rate of ultrafiltration would increase or more fluid passes to Bowman's capsule.

### Examiner's secrets

In 1(b) it is not sufficient to write that 'blood pressure increases'. You must be more specific in your answer and state that it increases at a specific point.

## Functioning of the kidney

### Checkpoints

1 Passive.
2 Glucose, amino acids, vitamins, some hormones, urea, uric acid, creatinine, ions, water.
3 Cells, platelets, plasma proteins which are too large to pass through the filter.
4 Longer.
5 The blood gains ions during the descending journey and loses them again from the ascending loop.
6 Base of loop.
7 It increases.
8 Detector – hypothalamus; coordinator – posterior lobe of pituitary (ADH); effector – collecting ducts of kidney.
9 Fall in water potential of the blood.

### Exam questions

1 Your essay should cover 10 of the following points in a logical sequence.
   (a)
   • There is a high pressure in the glomerular capillary.
   • The filter is provided by pores in the basement membrane of Bowman's capsule.
   • Particles of a certain size are allowed through, e.g. glucose, etc.
   • All the glucose is reabsorbed in the first convoluted tubule by diffusion and/or active transport.
   (b)
   • The loop of Henle enables the counter-current multiplier to operate.
   • Na⁺ ions are actively pumped out of the ascending limb into the medulla, creating a low water potential there.
   • Na⁺ diffuses into the descending limb.
   • The walls of the ascending limb are relatively impermeable to water.
   • Therefore reabsorption of water occurs from the permeable descending limb.
   • Water leaves the filtrate by osmosis.
   • The contents of the descending limb become more concentrated.
   • The maximum concentration occurs at the tip of the loop of Henle.

## Plant growth substances

### Checkpoints

1  There is a reduction in the numbers of fruits and seeds formed when environmental conditions are unfavourable.
2  Broad-leaved weeds have a large surface area and so receive a large dosage of weedkiller when applied; the narrow-leaved grasses do not. There may also be a difference in sensitivity of the different types of leaves.
3  The loss of soil nutrients to broad-leaved weeds and competition for light is reduced.
4  Some are more effective than naturally occurring substances, are easily manufactured, are stable in environmental conditions favourable for growth and can be more selective in the range of species affected.
5  The removal of the apical bud would allow the development of lateral buds. The addition of IAA would reinstate the inhibition.
6  Fruits produce ethene as they ripen. Ethene is a gas that diffuses through the air and causes ripening of unripe fruit.

### Exam question

Include all the following points in your answer.
• Auxins stimulate adventitious root growth.
• Auxins induce fruit development in the absence of pollination.
• Auxins can disrupt normal growth pattern making them useful as herbicides.
• Cytokinins mixed with auxins control differentiation of plants produced from tissue cultures.
• Anti-ageing properties of cytokinins used to keep cut flowers fresh.
• Cytokinins prolong the shelf life of fruit or vegetables.
• Giberelins influence fruit set.
• Giberelins break seed dormancy.
• Giberelins break bud dormancy in the spring.

## Effect of light on plant growth

### Checkpoints

1  In continuous light, the coleoptile splits open and the leaves emerge.
2  Curvature of the shoot towards the light puts the leaves in a good position for photosynthesis.
3  None.
4  To the left.
5  Natural sunlight contains more red light than far-red light.
6  It is the length of the dark period that is crucial.
7  Length of day.
8  A: dark period is short enough for long-day plants to flower but not long enough for short-day ones to do so.
   B: as there is no single long light period and yet long-day plants flower it must be the short dark period which is the critical factor inducing flowering.
   C: the dark period is long enough for short-day plants to flower but not short enough for the long-day ones to do so.
   D: the short light period in the middle of the dark period prevents short-day plants flowering as there is no single long dark period. Each dark period is, however, short enough to induce flowering in long-day plants.
9  Leaves.

### Exam questions

1  See definition in text.
2  Reference to long-day, short-day and day-neutral plants, together with an explanation of the link with phytochrome and how flowering is brought about in both long-day and short-day plants is required. Give named examples where possible.

## Nerves

### Checkpoints

1  | Hormone | Nervous |
   |---------|---------|
   | gland | sense receptor |
   | hormone | nervous impulse |
   | bloodstream | nerve fibre |
   | all over body | to a specific point |
   | target organ | effector |
   | usually slow | rapid |
   | widespread | localized |
   | long-lasting | usually brief |

2 In at dendrites, out through axon.

3 Action potentials jump from node to node at high speed.

4 Negative.

5 • Ion channels in the membrane open, $Na^+$ ion channels open first, $Na^+$ ions flow in generating an action potential.
   • $Na^+$ ion channels close again and $K^+$ ion channels open, $K^+$ ions flow out and the resting potential starts to be reformed.
   • $Na^+$ ions pumped out as $K^+$ ions pumped in; $K^+$ ions diffuse out rapidly but $Na^+$ ions diffuse back in only slowly creating the resting potential.

6 ATP.

7 Nodes of Ranvier and diameter of the axon.

8 The impulse can only move in a forward direction.

9 Provided the stimulus exceeds a certain threshold value an action potential results.

## Exam question

Your essay should include a well-labelled diagram and eight of the following points in a logical sequence.
• Diagram of motor neurone with cell body containing nucleus and with short dendrites and a long axon.
• Lipid myelin sheath surrounding axon.
• Interrupted at intervals by the nodes.
• At rest the outside of the axon membrane has a higher concentration of $Na^+$ ions than the inside.
• The reverse is true of $K^+$ ions.
• This results in a net positive charge of +70 mV on the outside.
• Stimulation causes an increase in membrane permeability.
• The arrival of the action potential opens $Na^+$ channels in the membrane protein.
• $Na^+$ ions rush in along the concentration gradient.
• The polarity reverses and the outside becomes relatively negative.
• The positive internal charge forces potassium out along its concentration gradient restoring a positive polarity to the outside.
• Sodium/potassium pump then re-establishes system ready for next impulse.
• Impulse travels as a wave of depolarization.
• It may jump from node to node of the myelin sheath.

## Synapses

### Checkpoints

1 Neurotransmitters.

2 Acetylcholine is broken down by cholinesterase and the components re-form on the other side of the synaptic cleft.

3 Facilitated diffusion.

4 ATP production and the generation of vesicles of transmitter substance.

5 The insecticide inhibits the enzyme which breaks down the neurotransmitter so new action potentials are propagated indefinitely and impulses merge.

6 Excitatory.

### Exam questions

1 (a) (i)  A, mitochondrion; B, synaptic vesicle.
     (ii) Acetylcholine / neurotransmitter.
   (b) The ions cause the synaptic vesicle to fuse with the presynaptic membrane and the contents are shed into the synaptic cleft.
   (c) (i)  Intrinsic proteins.
       (ii) The neurotransmitter combines with the protein (or the receptor site) causing the protein to change shape, thus opening the channels.
       (iii) The influx of sodium ions depolarizes the postsynaptic nerve.
   (d) The mitochondria provide large amounts of ATP which is used to pump the calcium ions out again or to resynthesize neurotransmitter after its breakdown.

## The central nervous system

### Checkpoints

1 Hypothalamus.

2 Cerebrum.

3 Contains more cell bodies.

4 An automatic response which follows a sensory stimulus.

5 A pathway of neurones involved in a reflex action.

6 The reaction is routine and predictable and does not need analysis and would be wasteful of the brain's capacity.

**Exam question**

- Cerebral hemispheres and their position.
- Correct location for intelligence, learning, language, speech, etc.
- As well as sensory interpretation.
- Motor areas controlling voluntary movement.
- Hypothalamus on the floor of the brain and its role in temperature regulation and osmoregulation and regulation of pituitary secretion.
- Correct position of cerebellum on lower rear area and its function of coordinating subconscious movement for balance, etc.
- Medulla in the brain stem as a relay centre for nerve tracts and its role in heartbeat, breathing, etc.
- Accurate mention of meninges and their function.

## Sensory receptors

### Checkpoints

1 Cornea and lens.
2 Retina.
3 Contains blood capillaries to provide food and oxygen.
4 With ageing the lens hardens and the ability to accommodate is lost, making it hard to focus on near objects. The reduction in pupil size (which occurs as light intensity increases) improves the depth of focus.

5

| Rods | Cones |
|---|---|
| numerous | less numerous |
| around periphery | in centre |
| one bipolar neurone | each served by own neurone |
| sensitive to low light | only stimulated by bright light |
| stimulated by most wavelengths | three types, each selectively responsive to different wavelengths |

6 Fovea.
7 ATP resynthesis.
8 Combined stimulation of cones.

**Exam question**

This question is about visual acuity. Groups of rod cells share a single nervous pathway to the brain so that the stimulation of more than one of a group still only produces a single impulse.

## Muscles

### Checkpoints

1 Both.
2 Actin only.
3 Where actin and myosin overlap the band appears dark.
4 Tropomyosin – fibrous; troponin – globular.

### Exam question

Your answer should cover 10 of the following points in a logical sequence.

- Banded into Z, A, I and H with a description or labelled diagram.
- Filaments are thick myosin and thin actin.
- H, myosin alone; I, actin alone; A, length of myosin.
- Diagram showing shortening of sarcomere with more overlap.
- A correct description of the changes in band width due to actin sliding over myosin, with no change in length of either.
- When the muscle is stimulated calcium ions are released.
- The protein troponin is activated with the calcium ions binding.
- This pushes another protein, tropomyosin, away from the attachment sites on the actin molecules.
- Myosin and actin can now be linked by cross-bridges.
- Show the clubbed head of the myosin molecule, which swings back and forth with the breaking and re-forming of bonds to produce a ratchet mechanism.
- This is energized by the hydrolysis of ATP.

# Microbiology

Recent developments in biotechnology have resulted in major advances in many different areas of medicine. Humans have studied microorganisms and manipulated their growth requirements in vessels called fermenters. Hormones, antibiotics and therapeutic proteins, such as monoclonal antibodies, are some of the cell products that have been mass produced and have been of benefit in the field of medicine. You also need to study the various control and prevention methods used against selected pathogenic organisms in addition to the natural defence mechanisms of the body. You need to appreciate the problems associated with the overuse of antibiotics.

## Exam themes

The growth requirements of microorganisms
Measurement of microbial population growth
The general structural features of a fermenter used for large-scale production
The advantages and disadvantages of batch and continuous culture
The use of monoclonal antibodies in medicine
The causes and means of transmission of named diseases
The social, economic and biological factors in the prevention and control of named diseases
The role of antibiotics in the treatment of infectious disease
The origin, maturation and mode of action of phagocytes and lymphocytes
The immune response
The actions of B lymphocytes and T lymphocytes in fighting infection
The distinction between different types of immunity
How vaccination can control disease

## Topic checklist

| ○ AS ● A2 | AQA/A | AQA/B | EDEXCEL | OCR | WJEC |
|---|---|---|---|---|---|
| Types of microbes | ○ | ● | ● | ● | ● |
| Growth and culture techniques | ○ | ● | ● | ● | ● |
| Fermentation | ○ | ● | ● | ● | ● |
| Pathogens and human disease | ○ | ● | ● | ○ | ● |
| Protection against disease | ○ | ● | ● | ○ | ● |
| Applications and contemporary issues | ○ | ● | ● | ● | ● |

# Types of microbes

It is important to study a range of microorganisms in order to appreciate their diversity with respect to size, complexity and reproduction. Therefore you need to consider bacteria, fungi and viruses and describe their general characteristics.

## Bacteria

Bacteria are the smallest cellular organisms and are prokaryotes. The kingdom includes both heterotrophic and autotrophic forms. The diagram below shows the structure of a typical bacterial cell.

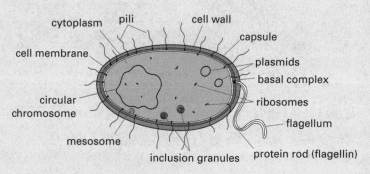

There are several ways of classifying bacteria. One way is by their shape. Bacteria may be:

→ rod-shaped or **bacilli**
→ spherical or **cocci**
→ corkscrew-shaped or **spirilla**
→ comma-shaped or **vibrio**

Further differentiation is often possible according to the way bacteria tend to be grouped, e.g. in pairs, chains or clusters. The filamentous bacteria are of particular interest because most are photosynthetic. This group, formerly called blue-green algae, is now known as cyanobacteria.

Another method of classification is based on the variation in nature of bacterial cell walls. The Gram stain technique distinguishes two types of bacteria.

→ *Gram-negative bacteria* have chemically complex walls where the glycoprotein is supplemented by large molecules of lipopolysaccharide and do not combine with dyes like gentian violet.
→ *Gram-positive bacteria* lack the lipopolysaccharide in their walls and do combine with gentian violet. These are more susceptible to antibiotics and the enzyme lysozyme, which occurs in human tears, than Gram-negative bacteria.

Bacteria reproduce asexually by binary fission, although certain bacteria can also reproduce sexually.

Although many bacteria are pathogens and cause deterioration to stored food, it is important to remember that many are beneficial to humans.

**Links**

Prokaryotes and eukaryotes are discussed on page 16.

**Checkpoint 1**

List the characteristics of a prokaryote such as a bacterium.

**Examiner's secrets**

The examiner will be impressed if you can refer to named microorganisms.

**Checkpoint 2**

What shape is a *Streptococcus* bacterium?

**Check the net**

You'll find more information on 'The microbial world' at helios.bto.ed.ac.uk/bto/microbes/microbes.html

**The jargon**

Lysozyme causes *lysis*, a dissolving of the walls of Gram-positive bacteria and so they swell by osmosis and burst.

**Links**

Different uses of bacteria can be found in various topics, such as the nitrogen cycle (page 60), and food production and manufacturing processes later in this section.

## Fungi

Although originally classed as plants, fungi are now recognized as a separate kingdom. This is based on the presence of the polysaccharide chitin, found in their cell walls, rather than the cellulose present in plant cell walls. Fungi also lack chlorophyll and are therefore unable to photosynthesize. Fungi are eukaryotic heterotrophs, generally feeding as saprobionts or parasites. The basic structural unit of a fungus is a thread-like multinucleate hypha. A collection of hyphae is called a mycelium. While some fungi cause damage and disease, particularly to plants, many are beneficial to humans, e.g. helping to maintain soil fertility by mineral recycling and decomposing organic matter, food and drink production and the production of antibiotics.

## Viruses

Viruses are extremely small and cannot be seen using the light microscope. They enter living cells and multiply with the assistance of the host cells. Viruses cause a variety of infectious diseases. Each virus particle, or virion, is made up of a core of nucleic acid surrounded by a protein coat, the capsid. Most viruses are found in animal cells and those attacking bacteria (bacteriophages) have the nucleic acid DNA. Other animal and plant viruses contain RNA. A widely studied virus is T2 phage, a bacteriophage, which infects *Escherichia coli*.

When a phage attacks a bacterium the capsid attaches to the surface of a host cell and the viral nucleic acid then enters. However, parasitism by phages is not typical of all viruses. In many cases the entire virus penetrates the host cell membrane and its nucleic acid passes to the cell nucleus.

**Retroviruses**, including HIV, contain RNA. Once inside the host, the RNA serves as a template for the formation of a single-stranded DNA. This molecule in turn acts as a template for the formation of a double-stranded DNA, which attaches to a host cell chromosome. Here, among other functions, it produces viral RNA for retroviruses of the next generation.

**Checkpoint 3**

What do you consider to be the most important difference between fungi and plants?

**Links**

Antibiotics are discussed on pages 122 and 123.

**Checkpoint 4**

For the replication of a virus in its host, why is it not essential for the protein coat of the virus to enter the host cell?

**Action point**

Draw a T2 phage and a tobacco mosaic virus.

**The jargon**

HIV, human immunodeficiency virus, causes the condition known as AIDS.

**Links**

Genetic engineering is discussed on pages 22 and 23.

**Exam questions**                    answers: page 124

1  (a) Complete and label the outline diagram below to show three features which may be found in bacteria but not in other groups of microorganisms.

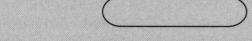

   (b) State two structural characteristics of fungi which bacteria do not possess.

(7 min)

# Growth and culture techniques

Microorganisms quickly reproduce given a suitable environment. In the laboratory they can be grown on a wide variety of substrates providing they are supplied with suitable physical conditions, nutrients and water. A variety of methods can be used to measure growth rate.

## Conditions necessary for growth

Microorganisms require the following conditions for growth.

→ *Nutrients*: in the laboratory nutrients are supplied in nutrient media and include carbon (usually organic such as glucose), nitrogen (organic and inorganic form) and growth factors such as vitamins and mineral salts.
→ *Temperature*: the range of 25–45 °C is favourable for the majority of bacteria as all growth is regulated by enzymes.
→ *pH*: most bacteria are favoured by slightly alkaline conditions (pH 7.4) whereas fungi prefer neutral to slightly acid conditions.
→ *Oxygen*: many microorganisms are aerobic and are termed 'obligate aerobes'. Some, while growing better in the presence of oxygen, can nevertheless survive in its absence; these are called 'facultative anaerobes'. Others cannot grow in the presence of oxygen and are called 'obligate anaerobes'.
→ *Water*: required for metabolic reactions.
→ *Osmotic factors*: to ensure the absorption of water, the water potential of the external environment must be higher (less negative) than the cell contents. Thus most microbes cannot grow in environments with a high solute concentration.

## Culture techniques

The growth of a culture of individual cells follows the same pattern as the growth of any population, i.e. lag phase, exponential phase, stationary phase and death phase.

Bacteria and fungi are cultured (grown) on, or in, media that are designed to supply the cell with all their nutritional requirements. Aseptic techniques, in which the apparatus and equipment are kept free of microorganisms, are used to prevent contamination of bacterial cultures and the surrounding environment (see upper diagram on page 115).

## Methods of measuring growth

There are several different methods of measuring growth.

→ Rough estimates of growth rates can be made by regularly measuring the diameter of a bacterial or fungal colony as it spreads from a central point to cover the surface of a solid growth medium.
→ Unit samples from broth cultures can be filtered, dried and weighed.
→ Changes in cell density can be measured by the following methods.
  → *Haemocytometer* measures total counts (both living and dead cells).

**Links**

Enzymes and temperature are discussed on page 12.

**Action point**

Explain, in biological terms, how most food preservation methods use the principle of removing conditions necessary for growth.

**Checkpoint 1**

What changes do you think may occur in cells adapting to a new culture medium?

**Links**

The sigmoid growth curve for population is looked at on page 62.

**Examiner's secrets**

This section of the topic consists almost entirely of practical work. An examiner will be impressed if you can incorporate your practical experience into your exam answers and be aware of the safety issues and potential hazards associated with microbiology.

→ *Serial dilution* and plating onto a solid nutrient medium (see lower diagram, this page) allows viable counts (living cells only) to be made by counting the number of colonies. Here it is assumed that one cell divides to form a colony.

→ *Photometry*: a colorimeter can be used to measure light intensity and turbidity (cloudiness) as cell numbers increase.

**Checkpoint 2**

What advantage may be gained from choosing to plot the log of cell number counted in a culture over a period of time?

**Checkpoint 3**

By what process does one cell give rise to a colony?

**Checkpoint 4**

What method would a bacteriologist use to estimate the changes in the numbers of viable cells in a culture over a period of time?

**Checkpoint 5**

Why is it necessary to sterilize at a temperature of 121 °C for 15 min?

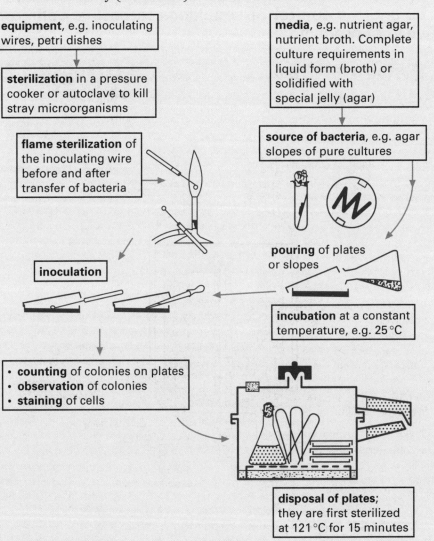

**equipment**, e.g. inoculating wires, petri dishes

**sterilization** in a pressure cooker or autoclave to kill stray microorganisms

**flame sterilization** of the inoculating wire before and after transfer of bacteria

**media**, e.g. nutrient agar, nutrient broth. Complete culture requirements in liquid form (broth) or solidified with special jelly (agar)

**source of bacteria**, e.g. agar slopes of pure cultures

**pouring** of plates or slopes

**inoculation**

**incubation** at a constant temperature, e.g. 25 °C

- **counting** of colonies on plates
- **observation** of colonies
- **staining** of cells

**disposal of plates**; they are first sterilized at 121 °C for 15 minutes

*transfer* 1 cm³   1 cm³ 1 cm³ 1 cm³ 1 cm³

original culture

tubes containing 9 cm³ of sterile saline

**serial dilutions** are carried out so that each colony that grows, when a sample is plated, represents a single cell or spore in the original sample

| concentration | 1 | $10^{-1}$ | $10^{-2}$ | $10^{-3}$ | $10^{-4}$ | $10^{-5}$ |
|---|---|---|---|---|---|---|
| dilution factor | 0 | 10 | 100 | 1000 | 10,000 | 100,000 |

**Exam question**                                   answer: page 124

Describe and explain the procedure that you would use to safely obtain a pure culture of one species of bacterium from a tube containing a mixed bacterial suspension in nutrient medium. (15 min)

# Fermentation

Biotechnology may be defined as the industrial application of biological processes. For many of these processes the key steps are conducted in a fermenter vessel. You should understand the difference between batch and continuous fermentation and how industrial processes involve the need for aseptic entry of material and the control of conditions to ensure optimum yield of product.

## Fermenter design

A sterile closed fermenter vessel excludes the possibility of contamination by other organisms (see diagram below).

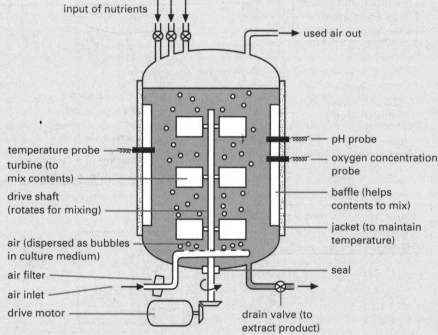

input of nutrients

used air out

temperature probe
turbine (to mix contents)
drive shaft (rotates for mixing)
air (dispersed as bubbles in culture medium)
air filter
air inlet
drive motor

pH probe
oxygen concentration probe
baffle (helps contents to mix)
jacket (to maintain temperature)
seal
drain valve (to extract product)

The products of fermentation may be:

→ a dissolved substance that accumulates in the medium, e.g. the antibiotic penicillin is produced by the fungus *Penicillium*; in this case the used culture medium, after filtering, is kept and processed
→ the microorganism itself, e.g. single-cell protein (mycoprotein) manufacture; in this case the medium may be recycled and the solid processed

## Batch versus continuous cultivation

In batch fermentation, as the name suggests, for a particular reaction to occur, growth is allowed to continue up to a specific point at which the fermenter is emptied and the product extracted. The fermenter has to be cleaned and sterilized in readiness for the next batch.

In continuous fermentation, once the fermenter is set up the used medium and products are continuously removed. The raw materials are added throughout.

**Checkpoint 1**

Define fermentation.

**Checkpoint 2**

Why is fermentation a misnomer?

**Checkpoint 3**

What are the advantages and disadvantages of single-cell protein production?

**Checkpoint 4**

What is the purpose of fixed aeration?

## Antibiotic production

For the commercial production of penicillin, the fungus *Penicillium notatum* (or *P.chrysogenum*) is grown in a batch culture in a fermenter. The antibiotic is secreted into the culture solution by the fungus after the initial growth phase is over and when glucose is depleted. The diagram below shows the sequence of processes involved in the commerical production of penicillin.

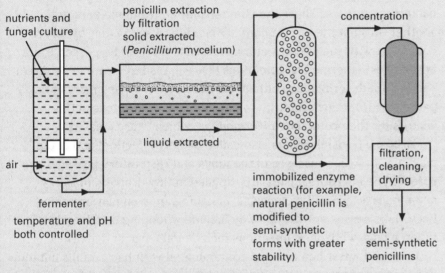

Antibiotic production is a secondary metabolism, i.e. at a period in the life of the fungus when there is a change away from its optimum conditions. The primary metabolism is the norm, when the fungus is breaking down glucose to release energy and producing an increase in its own biomass (see graph below).

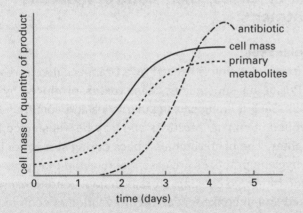

**Checkpoint 5**

What are the advantages and disadvantages of continuous fermentation?

**Checkpoint 6**

What is the importance of routine measurements of pH, temperature and oxygen concentration in a fermenter producing penicillin?

**Checkpoint 7**

Suggest how *Penicillium* when free-living uses penicillin when food sources are depleted.

**Links**

See also immobilized enzymes on page 13 and antibiotics on pages 122–3.

**Exam questions**                          answers: pages 124–5

1  (a) With reference to the diagram above explain the importance of each of the following to the industrial manufacture of penicillin.
   (i)   The cooling jacket.
   (ii)  The pH probe.
   (iii) The air filter.
   (iv)  The inclusion of ammonium as one of the nutrients
   (b) Suggest why it is often cheaper to use living organisms than conventional laboratory techniques to produce chemicals.

                                                          (15 min)

# Pathogens and human disease

Another aspect of microbiology is the effect of the different pathogenic organisms and their control. Bacteria, viruses and fungi can be pathogens of humans, their crops and their domestic animals and the consequences can be of great economic significance.

## Disease transmission ●●●

Many of the important diseases in humans are caused by **pathogenic** bacteria or viruses. These may be transmitted from one person to another through the air, via contaminated food and water, by direct contact or with the aid of vectors. A disease that can be passed from host to host is termed an **infectious** disease. Diseases spread by direct contact are described as **contagious**. Vectors are organisms that carry pathogens from one organism to another, e.g. mosquitoes act as vectors for the protoctistan *Plasmodium*, which causes malaria.

Airborne transmission is also called droplet infection, and is the main way in which diseases of the lungs and respiratory passages are spread. The microorganisms are expelled in tiny droplets of saliva and mucus as a result of sneezing, coughing, or even just breathing. Bacterial diseases spread this way include whooping cough and tuberculosis, while viral infections include influenza and measles.

Food and water can become contaminated with bacteria if sanitation and hygiene are poor. *Salmonella* food poisoning is spread in or from the undercooked meat of infected animals, while diseases such as cholera and typhoid are mainly spread through the contamination of water by the untreated faeces of infected individuals.

## Selected diseases, their control methods and treatment ●●●

### Bacterial diseases

*Salmonella* food poisoning is caused by a Gram-negative rod-shaped bacterium. Poisonous substances, called **toxins**, produced by the bacteria affect the gut lining causing diarrhoea and vomiting. It is commonly found in animal intestines and may be transferred to meat at animal slaughter. The bacterium multiplies during storage of infected food. Good hygiene practices are essential to avoid food poisoning. These include thorough cooking, storage in cool conditions, separation of cooked and uncooked meat, the prevention of contamination from carriers. Analysis of antigenic types may enable the tracing of the source of an infection.

Cholera is caused by a Gram-negative bacterium, endemic in some areas of the world. Toxins affect the gut lining causing watery diarrhoea leading to severe dehydration and possibly death. Humans can act as a reservoir and as carriers. This disease is often spread by infected water in which the organism does not multiply but may become easily widespread. Protective measures are a clean water supply, the safe disposal of sewage and the hygienic preparation of food. Powerful antibiotic therapy is possible but treatment is largely by rehydration.

**Links**

Parasites are discussed on pages 52 and 53.

**Examiner's secrets**

Check which diseases you are required to study for your specification.

**Checkpoint 1**

Why is it important to allow a frozen chicken to thaw completely before cooking?

**Action point**

List the methods that should be used to prevent microbial contamination of foods.

**The jargon**

An *endemic* disease is always present at low levels in an area.
A *carrier* is infected by a disease organism, shows no symptoms, but can pass the disease on to another individual.
The *reservoir* is the place where a pathogen is normally found.

118

## Protoctistan disease

Malaria is a major human disease, caused by the parasite *Plasmodium*. The life cycle occurs in the human body and the female mosquito acts as a secondary host and vector. It is endemic in some subtropical regions. The infection is acquired by a human when an adult female mosquito, already infected with the malarial parasite, inserts its mouthparts into the individual's skin. The parasite infects the liver, where it multiplies, and then the red blood cells, which eventually burst to release parasites, causing severe bouts of fever. Infected human or animal blood is taken in by a feeding mosquito, which transmits the parasite to a new host.

Prevention relies on a knowledge of the vector and the life cycle of both the vector and parasite in order to exploit their weak points. A variety of methods are used either to prevent transmission (e.g. prevent biting by use of nets, clothing, insect repellents) or to control the life cycle of the vector by the use of insecticides, biological control, sterilization of males to reduce breeding, drainage of swamps, etc.

## Virus diseases

Several highly significant human diseases are caused by viruses, including rubella, shingles, measles, poliomyelitis, influenza and HIV. Influenza is caused by a tiny virus consisting of a central strand of RNA, coated in protein, surrounded by an outer lipoprotein coat. It infects cells of the lining of the upper respiratory tract causing sore throat, cough and feverishness. Transmission is by droplet infection. In general, antibiotic treatment has no effect and therefore symptoms are treated as necessary. Vaccines are available but due to the number of antigenic types (strains) and the regular emergence of new ones, they are not always effective.

AIDS, the acquired immune deficiency syndrome, is caused by a retrovirus known as human immunodeficiency virus (HIV) which damages the immune system. People with AIDS are highly susceptible to 'opportunist diseases', infections and cancers that take advantage of an immune system in collapse.

Within the host, retroviruses cause the host cells to translate the viral RNA into DNA which is then incorporated into the host cell genome. In this provirus form the viral genome directs the production of new virus particles. HIV infects helper T-cells which are essential to cell mediated immunity and so the body's immune system is rendered ineffective. Once infected with HIV an individual is said to be HIV positive. As the virus may be dormant for up to eight years, HIV positive individuals do not suffer any symptoms during this period but can act as carriers. Transmission of HIV requires the transfer of body fluids, such as semen or blood, containing infected cells. Preventative precautions such as restricting the number of sexual partners, using a condom during sexual intercourse, and encouraging intravenous drug users not to share a needle are the best means of containing the disease at present. Also, the drug AZT inhibits replication of the AIDS virus and can slow the process of the disease but at present there is no known cure.

**Checkpoint 2**

Why does the malarial parasite need two hosts?

**Action point**

Discuss the advantages and disadvantages of the various control methods for malaria.

**Checkpoint 3**

State one similarity between the malarial parasite and HIV.

**Checkpoint 4**

Explain why HIV is described as a parasite.

**Action point**

Summarize the information about the four diseases in the spread by constructing a table with the headings – name of organism, source, method of transmission, prevention/control.

**Exam question**                                   answer: page 125

List the main groups of microorganisms that include pathogens of medical or agricultural importance. Give an example of an organism in each group, briefly indicating how the disease it causes can be controlled. (15 min)

119

# Protection against disease

The human body operates a number of defence mechanisms against disease, e.g. natural barriers such as the skin. Resistance to disease also depends on the general health and diet. However, in this spread you need to study the role of the immune system, which involves the recognition of foreign material and the production of antibodies, which help to destroy it. It is also possible to induce an individual to produce antibodies without them suffering from the disease. This is the basis of immunization or vaccination.

**Don't forget**

The body possesses a number of other defence mechanisms against disease apart from the immune system.

## The immune system

Immunity is the ability of an organism to resist disease. An **antigen** is any substance that when introduced into the blood or tissues induces the formation of antibodies, or reacts with them if they are already present. An antibody is a substance produced by lymphocytes (white blood cells), in the presence of a specific antigen, that can combine with that antigen to neutralize, inhibit or destroy it.

## Immunity

Immunity involves two initial processes:

→ the recognition of foreign material when it enters the body and the mobilization of cells and antibodies capable of removing the foreign material, or antigens, quickly and effectively
→ the formation of memory cells, which carry out the secondary immune response

There are two systems of immunity in mammals: a cell-mediated immune response and a humoral immune response. Two types of lymphocytes are involved, both of which have receptor sites on their membranes for recognizing antigens. Both types are derived from stem cells in the bone marrow. The lymphocytes that migrate to the thymus gland for maturation are called **T-lymphocytes** or T-cells, whereas those which remain in the bone marrow to complete their development are called **B-lymphocytes** or B-cells.

**Checkpoint 1**

After production in the bone marrow where do T-lymphocytes complete their development?

## Cell-mediated immune response

The thymus gland is active from birth until a mammal is weaned. During this time it causes lymphocytes to mature and become 'immunologically competent', and capable of synthesizing new receptor molecules and incorporating them into the plasma membrane. T-lymphocytes leave the thymus and circulate in the blood and body fluids. If a T-lymphocyte meets an antigen for which it has a receptor site, it is stimulated to divide many times by mitosis, forming a clone. Each cell in the clone can attach to a complementary antigen and destroy it. Among these dividing cells three functional sub-populations occur (in addition to the formation of memory cells):

**Checkpoint 2**

Name one group of microorganisms against which cell-mediated immunity is *most* effective.

**Checkpoint 3**

In cell-mediated immunity name the cell first activated in response to a pathogen.

→ T killer cells cause lysis of the target cells, e.g. they destroy virus infected or cancer cells.
→ Helper T cells which activate B lymphocytes to initiate an antibody response.

→ T suppressor cells suppress other cells in the immune system, e.g. by turning off antibody production when the antigen is no longer present.

## Humoral immune response

There are many thousands of different B-lymphocytes, each with just one type of receptor on its cell surface. If an antigen is recognized, the B-lymphocytes are stimulated to divide by mitosis forming a clone of plasma cells in the lymph node, and also forming memory cells. The plasma cells live only a few days but can synthesize and secrete vast quantities of specific antibody molecules.

The memory cells persist for a long time, sometimes for life. They are responsible for the secondary immune response and confer active immunity against the specific antigen. If that antigen is encountered again, the memory cells can recognize it and stimulate the immediate production of massive quantities of antibody.

## Types of immunity

There are three types of immunity.

→ *Hereditary immunity* occurs passively due to the inheritance of genes for disease resistance.
→ *Naturally acquired immunity* may occur passively or actively.
  → Natural *passive* immunity may be due to the transfer of antibodies from mother to fetus across the placenta or from mother to newborn offspring via colostrum, the first secretion of the mammary gland. The immunity is only temporary since no memory cells have been formed.
  → Natural *active* immunity is achieved as a result of exposure to infection. The body manufactures its own antibodies in response to the presence of antigens on the infectious agent and also forms specific memory cells. If the same agent is encountered again, it can be flooded with antibodies and eliminated before it causes disease.
→ *Artificially acquired immunity* also occurs passively or actively.
  → Artificial *passive* immunity results from the injection of ready-made antibodies into the body, and again, since there are no memory cells it is only temporary. It is useful as a preventative measure for diseases that are difficult to immunize against, such as tetanus and diphtheria. It may also be used as a treatment for certain diseases, such as rabies, where infection has already occurred which is too dangerous to leave to the body's natural immune system.
  → Artificial *active* immunity is achieved by injecting vaccine into a healthy individual. The body is stimulated to produce antibodies and memory cells against the antigen in the vaccine, and thus acquires immunity to subsequent infection by that disease organism.

### Exam question

answer: page 125

Describe what is meant by 'antigen' and 'antibody'. Explain how infection with a disease organism can often result in lifetime protection from a second attack. (15 min)

**Checkpoint 4**

Name one group of microorganisms against which humoral immunity is *most* effective.

**Checkpoint 5**

In humoral immunity name the cell first activated in response to a pathogen.

**Checkpoint 6**

Why is passive immunity temporary?

**Links**

The effectiveness of vaccination programmes is dealt with on page 123.

**Test yourself**

Distinguish between the following terms:
antigen and antibody;
active and passive immunity;
B lymphocytes and T lymphocytes;
infectious and contagious.

# Applications and contemporary issues

Recent advances in biotechnology have brought about the discovery of a variety of new ways of treating disease. Monoclonal antibodies are used to diagnose disease, to locate and isolate other molecules and in diagnostic testing. Fermenters have enabled the large-scale production of genetically engineered products such as insulin, mycoprotein and antibiotics. You need to consider the mode of action of the latter, their effects on microorganisms that cause disease and the problems associated with their overuse.

## Monoclonal antibodies

Traditional methods of producing antibodies involving laboratory animals have resulted in impure preparations. Natural antibodies are a mixture because several antibodies may be formed in response to an invading bacterium. A monoclonal antibody is one that responds to only one foreign antigen. It is the product of cells called hybridomas, made from mice myeloma cells. In monoclonal antibody production, spleen cells are exposed to a single known antigen. They produce a single specific antibody to it. These spleen cells are fused with myeloma cells and the result is a clone producing a specific antibody. The hybridoma cells are grown by tissue culture and the antibody they secrete is collected and concentrated.

Monoclonal antibodies have a number of uses:

→ to treat a range of infections
→ in the diagnosis of disease
→ to separate a particular chemical from a complex mixture
→ antibodies may be tagged with a marker the detection of which is used for the diagnosis of specific conditions, e.g. pregnancy-testing kits
→ to act as 'magic bullets' – linking anti-cancer drugs with monoclonal antibodies attracted to cancer cells so that drugs can be delivered to specific target cells, such as tumours, so improving effectiveness and reducing side-effects

### Pregnancy testing

Home pregnancy-testing kits make use of immobilized antibodies on a urine dipstick to detect traces of human chorionic gonadotrophin (hCG), a hormone released from the placenta. Antibodies 'tagged' with blue latex combine with the hormone to produce a readily visible result.

## Antibiotics

Antibiotics are substances that in low concentration inhibit the growth of microorganisms. Antibiotics effective against a wide range of pathogenic bacteria are called 'broad-spectrum' antibiotics, e.g. chloramphenicol, tetracyclines. Others, e.g. penicillin and

**The jargon**

A *myeloma* is a type of cancer that is associated with the abnormal production of irregular antibodies. It occurs in antibody-producing B-cells that have lost their normal controls.

**Links**

Production of antibiotics by fermentation is discussed on pages 116–17.

streptomycin, are 'narrow-spectrum' antibiotics as they are effective against only a few pathogens. Antibiotics affect the cell walls and membranes of microorganisms and also affect nucleic acid and protein synthesis.

## Antibiotic resistance

The use of antibiotics to treat disease in the human population has had obvious benefits. However, in agriculture, although the prophylactic use of small quantities of antibiotics in animal feed has led to healthier, faster growing animals, this widespread increased use of antibiotics has led to the development of resistance among many species of bacteria. Resistance arises by mutations occurring randomly within populations of organisms, which then confer an advantage in the presence of that antibiotic. There is evidence that resistance may also be passed from one organism to another on plasmids, during conjugation (sexual reproduction). Some bacteria have developed an enzyme, penicillinase, which renders penicillin ineffective. It is also possible for some bacteria, e.g. *Staphylococcus aureus*, to develop resistance to several antibiotics. Such organisms are becoming a more common problem and the antibiotics available to treat them very limited in number.

## The effectiveness of vaccination programmes ●●●

Vaccination has made a major contribution to reducing the incidence of certain diseases in the population. Its effectiveness is improved the greater the proportion of the population vaccinated so that the incidence is reduced until eventually the disease is eradicated locally and then possibly globally, as in the case of smallpox, with international cooperation. Smallpox vaccination has been effective because the organism does not have a high mutation rate so the vaccine remains effective, unlike the influenza organism which mutates frequently and causes periodic epidemics due to lack of resistance in the population and lack of effective long-term vaccination.

Other considerations are:

→ contraindications against vaccines
→ global travel increases the risk of the spread of infections

**The jargon**

*Prophylactic,* in this context, means applying antibiotics to *prevent* infection.

**Checkpoint 1**

Explain how the use of antibiotics in animal feed leads to the development of resistance of bacteria in humans.

**The jargon**

*Resistance* to a chemical poison is the ability of an organism to survive exposure to a dose of that poison which would normally be lethal to it.

**The jargon**

An *epidemic* is where there is a significant increase in the usual number of cases of a disease often associated with a rapid spread.

**Exam question**                                    answer: page 126

Describe the production and therapeutic applications of monoclonal antibodies. (15 min)

**Examiner's secrets**

Therapeutic means the cure of infections after they have been diagnosed, so limit your answer accordingly.

# Answers
## Microbiology

## Types of microbes

### Checkpoints

1 No distinct nucleus; DNA not incorporated in chromosomes but made up of a single circular strand; no spindle forms at cell division; membrane-bounded organelles are absent.
2 Spherical.
3 Chitin cell wall in fungi, cellulose in plants; no chlorophyll in fungi.
4 As only the viral DNA enters the bacterial cell and the protein coat remains on the outside, it is the nucleic acid of the virus that carries the information to produce the protein coat. Inside the host cell fresh 'coat' proteins are made on the host cell's ribosomes.

### Exam questions

1 (a) Completed diagram labelled with three of the following: cell wall containing murein/no cellulose or chitin present; circular DNA lying free in the cytoplasm; no nuclear membrane present; mesosomes; flagellae; mucilaginous capsule; pili.
   (b) Cell wall contains chitin; consist of hyphae; form a mycelium; no distinct cell boundaries/multinucleate; multicellular; sporangia.

## Growth and culture techniques

### Checkpoints

1 During the lag phase hydrolytic enzymes specific to the food source are produced.
2 The curve of exponential growth, when presented as a log plot, becomes a straight line, making interpretation easier.
3 Asexual reproduction.
4 Dilution plate method since only live cells give rise to a fresh colony on a dilution plate.
5 Gram-positive bacteria, e.g. *Clostridium*, produce endospores which are resistant to boiling.

### Exam question

Your answer should include 10 of the following points in a logical sequence.
- Aseptic or sterile technique used.
- Inoculating loop or Pasteur pipette sterilized in the flame of a Bunsen burner or an autoclave or irradiation.
- Remove cotton wool plug or stopper and flame neck of culture bottle.
- Streak out with a sterile loop or spread with a sterile spreader or drops on an agar plate.
- Stress that plates are sterile.
- Ensure lid of petri dish is exposed to air for a minimum period of time.
- Seal plates with tape.
- Incubate at a stated temperature (25 °C or below) for one to two days.
- Select an isolated colony.
- Swab work area to maintain aseptic technique.
- Explanation of asexual reproduction by isolated colony.
- Stress the importance of a temperature below 25 °C to prevent the development of pathogenic bacteria.
- Stress the reason for sealing the plate to prevent contamination into the air.
- The correct disposal of plates using an autoclave.

### Examiner's secrets

This question should prove straightforward if you have carried out the procedure in the laboratory. It demonstrates the importance of revising practical work. The question requires you to describe the technique of subculturing. Note how the word 'safely' requires you to mention points such as not leaving petri dishes open but sealing them; inoculating equipment should be sterilized before *and* after use; sterilize work area; safe disposal of plates at end of procedure. A well-annotated diagram could be used to gain many of the marks. Make every effort to show the examiner that you have had experience of practical work.

## Fermentation

### Checkpoints

1 The biochemical process involving the culture of microorganisms in an aqueous suspension in a culture vessel known as a fermenter.
2 Strictly speaking fermentation means the anaerobic breakdown of glucose to ethanol or lactic acid. Nowadays the word is taken to include aerobic processes.
3 A much faster growth rate is achieved compared with agricultural products, with doubling time measured in hours; a wide range of waste materials, e.g. cheese whey, can be used as substrates. It is a healthy food high in protein, high in fibre and low in cholesterol.
4 For maximum growth of aerobes, also helps to mix the culture to improve contact with nutrients (with the help of a stirrer).
5 Advantages: quicker, no need to empty, clean and refill; by adjusting addition of nutrients the rate of growth can be maintained at a constant level to provide maximum yield. Disadvantages: only suited to the production of biomass or metabolites associated with growth; a sophisticated monitoring technology is required.
6 Monitoring the process of product synthesis; checking that the culture is not contaminated by other organisms causing alternative reactions to occur.
7 To destroy competitors which may be competing for the same food source.

### Exam questions

1 (a) (i) *Penicillium* generates heat during its respiration which would denature its own enzymes; the jacket maintains the optimum constant temperature for enzyme activity.

(ii) It adjusts or maintains a constant or optimum pH; if the pH is too high the rate of reaction will slow due to the inactivation of enzymes.

(iii) It removes or prevents the contamination by airborne microorganisms, which would reduce the production of antibiotic due to the introduction of toxins.

(iv) Ammonium is a source of nitrogen for the production of amino acids and proteins by the microorganism.

(b) As microorganisms contain all the necessary enzymes to carry out complex chemical reactions, the processes are carried out at lower temperature with less electrical energy needed and so costs in the long term are reduced. The enzymes are also a renewable source.

## Pathogens and human disease

### Checkpoints

1 The recommended cooking time would be insufficient to kill bacteria if any part remained frozen.
2 Since the parasites are the target of specific immune attack in the primary host, they take temporary refuge by invading a secondary host.
3 Both invade humans; both invade the blood system; both reinfect more blood cells.
4 It lives in another living organism and causes it harm.

### Exam question

Your answer should include a range of microbes with the appropriate control measures.
Protoctista: malarial parasite (*Plasmodium*) or other suitable example.
Fungi: ringworm, athlete's foot, rust, mildew.
Bacteria: tetanus, *Salmonella*, cholera, or any named pathogen.
Virus: HIV, influenza, poliomyelitis, rubella, tobacco mosaic virus, etc.

- Some discussion on prevention being more desirable than the patient actually suffering the disease, stressing the economic advantages of preventative medicine.
- Control of bacterial and viral diseases by vaccination to induce active immunity.
- Vaccination programmes can be very successful, e.g. smallpox, stressing that a high proportion of the population possess immunity.
- The prevention of disease by barriers to infection, e.g. various named measures to exclude mosquitoes including the isolation of sufferers in quarantine.
- Antibiotics to control bacterial disease.
- A discussion of the problem of overuse of antibiotics resulting in resistance.
- Named example of resistance.
- Food hygiene.
- Example of organism causing food poisoning.
- Food inspection.
- Food preservation methods.
- Destruction of vectors, e.g. by insecticide use against the mosquito.
- A problem of insecticide resistance.
- Methods of killing mosquito, e.g. oil, biological control, drainage of swamps.

## Protection against disease

### Checkpoints

1 Thymus gland.
2 Viruses.
3 T-lymphocyte.
4 Bacteria.
5 Memory cell/B-lymphocyte.
6 No memory cells are formed.

### Exam question

Your answer should include the following points and must contain the definitions of antigens and antibodies.

- Antigens are substances which trigger the immune response. They are known as self antigens if they belong to the host and non-self antigens if they belong to a foreign source.
- Antibodies are specific proteins, known as immunoglobulins.
- They are produced by the B-lymphocytes.
- When their receptors detect an antigen the B-cells which carry the correct antibody for the introduced antigen multiply rapidly to produce a clone.
- Some of these become antibody secretors, known as plasma cells.
- Others remain in the circulation as memory cells, but survive in the body ready to clone immediately there is any future invasion by the same antigen.
- A second invasion produces a more rapid response compared with the first.

# Applications and contemporary issues

## Checkpoints

**1** It is thought that antibiotic residues may accumulate in the food chain when humans eat meat. This overuse of antibiotics has greatly accelerated the spread among bacteria of resistance to antibiotics. The development of resistance occurs as follows. In a large enough population of bacteria a small number are naturally resistant through random genetic variation. Exposure to antibiotic eliminates the susceptible majority. The small number of resistant survivors multiply rapidly in the absence of competition. The population becomes more or less totally antibiotic-resistant.

## Exam question

The production part of the question is detailed in the diagram opposite. Therapeutic applications should include some detail of the following.

- Purification of important chemicals, such as interferon, to remove organic contaminants.
- Combined with cancer treatment to target the surface of cancer cells. This site-specific release of toxins destroys the cancer cells without harming healthy tissue.
- Prevention of rejection of transplants.

### Examiner's secrets

7 marks are allocated for your answer to the part of the question about the production of monoclonal antibodies. It must involve a description but well-annotated diagrams would be acceptable. Although monoclonal antibodies are used in the diagnosis of disease, tissue typing in transplantation, pregnancy-testing kits, etc., the question does not require you to discuss these points. You must confine your answer to their use *after* diagnosis.

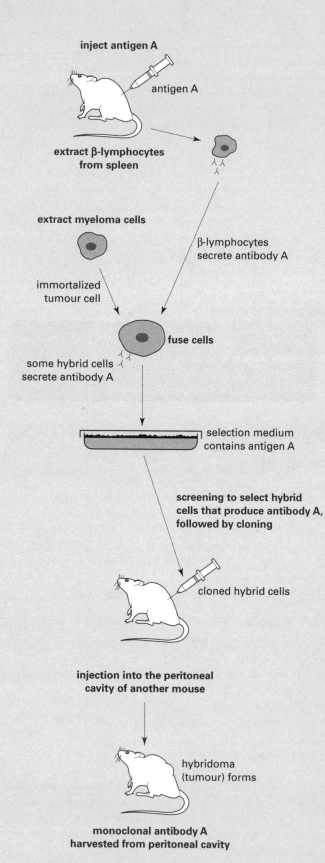

inject antigen A

antigen A

extract β-lymphocytes from spleen

extract myeloma cells

β-lymphocytes secrete antibody A

immortalized tumour cell

fuse cells

some hybrid cells secrete antibody A

selection medium contains antigen A

screening to select hybrid cells that produce antibody A, followed by cloning

cloned hybrid cells

injection into the peritoneal cavity of another mouse

hybridoma (tumour) forms

monoclonal antibody A harvested from peritoneal cavity

# Reproduction and genetics

In addition to increasing numbers, sexual reproduction produces offspring that are genetically different and so is a means of increasing variety, enabling plants and animals to adapt to changing conditions. Flowering plants have become the dominant plant group on Earth and humans the dominant animal group. Although humans have bred plants and animals for thousands of years, it is only in recent times that they have begun to understand the underlying genetic principles. This has enabled genetic engineers to manipulate human DNA as well as that of animals and crop plants of economic importance. You should be aware of the potential hazards of this course of action.

## Exam themes

Gamete development in flowering plants
The structural changes which occur after fertilization in plants
Describe and explain human gametogenesis
The role of hormones in the menstrual cycle, pregnancy, birth and lactation
The behaviour of chromosomes during meiosis
How meiosis and fertilization can lead to variation
The use of genetic diagrams to solve problems involving monohybrid and dihybrid crosses
The importance of polyploidy in agriculture
How genetic screening is carried out, its advantages and disadvantages and the need for genetic counselling
The theoretical basis of genetic fingerprinting
The cloning of plants from tissue culture
The advantages and disadvantages of cloning

## Topic checklist

O AS ● A2

| | AQA/A | AQA/B | EDEXCEL | OCR | WJEC |
|---|---|---|---|---|---|
| Flower structure and gamete development | | | O | ● | ● |
| Pollination and fertilization | | | O | ● | ● |
| Seed development and germination | | | O | ● | ● |
| Human reproduction | O● | ● | O | ● | ● |
| Hormonal control of reproduction | O | ● | O | ● | ● |
| Fertilization and development | O● | ● | O | ● | ● |
| Variation | ● | ● | ● | ● | ● |
| Meiosis | O● | O● | O● | O● | O |
| Mendel's laws | ● | ● | ● | ● | |
| Deviations from Mendel's laws and sex linkage | ● | ● | ● | ● | |
| Mutations | ● | O● | ● | ● | ● |
| Genetic counselling and gene therapy | O | O | ● | ● | O |
| Applications of reproduction and genetics | O | O | O● | ● | ● |

# Flower structure and gamete development

The flowering plants or angiosperms are the most successful of all terrestrial plants. The flower is the organ of reproduction and is usually hermaphrodite. In angiosperms the female part, the ovule, is never exposed but is enclosed within a modified leaf, the carpel. You will need to study the structure of a typical flower as well as how it produces the male and female gametes, pollen and ovules.

## Flower structure

A typical flower is made up of four sets of modified leaves, each set being referred to as a whorl.

→ *Calyx* (outer whorl): sepals, which protect the flower in bud.
→ *Corolla*: petals, brightly coloured to attract insects.
→ *Androecium*: stamens, which comprise a long filament at the end of which are the anthers which produce pollen grains containing the male gamete. As well as supporting the anther the filament contains vascular tissue that transports food materials necessary for the formation of pollen grains. The anther is usually made up of four pollen sacs arranged in two pairs, side by side.
→ *Gynoecium* (inner whorl): made up of one or more carpels. Each carpel is a closed structure inside which one or more ovules develop. Each ovule develops from the nucellus and becomes surrounded by two integuments. The lower part of the carpel, which surrounds the ovules, is called the ovary and bears at its apex a stalk-like structure, the style. This ends in a receptive surface, the stigma. The diagram below shows the structure of a typical insect-pollinated flower.

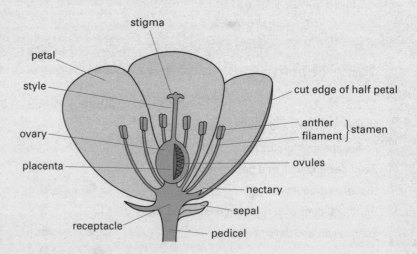

## Pollen grain development

In the early stages of development, each pollen sac contains a mass of microspore mother cells, surrounded by a nutritive layer called the tapetum (see diagram opposite).

### The jargon

A *hermaphrodite* is an organism with both male and female organs.

### Check the net

You'll find some good exercises on flower structure at wsuonline.weber.edu/course.botany.130/unit1_la.htm

### Checkpoint 1

How do flowers attract insects?

### Checkpoint 2

What is the nutritional value of nectar to insects?

### Checkpoint 3

Describe the adaptations of pollen grains to insect pollination.

### Action point

Annotate the diagram of a typical flower describing the functions of the various parts.

### Checkpoint 4

What is the function of the tapetum layer in the pollen sac?

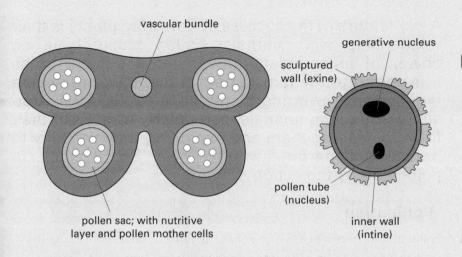

vascular bundle

generative nucleus

sculptured
wall (exine)

pollen tube
(nucleus)

inner wall
(intine)

pollen sac; with nutritive
layer and pollen mother cells

Each microspore mother cell undergoes **meiosis** to give a tetrad (four) of haploid cells. Each of these cells develops into a pollen grain. A thick outer wall, the **exine**, and a thin inner wall, the **intine**, are secreted. Inside the pollen grain the haploid nucleus undergoes mitosis to produce a generative nucleus and a pollen tube nucleus. The generative nucleus will later give rise to the two male nuclei. As the pollen grains mature, changes take place in the anthers. The cells of the tapetum shrink and degenerate, and fibrous layers develop beneath the epidermis. When the pollen is ripe, the outer layers of the anthers dry out and tensions are set up in lateral grooves. Eventually dehiscence occurs and the edges of the pollen sacs curl away exposing the pollen grains.

## Ovule development ●●●

In the early stages of the development of an ovule, the nucellus arises as a tiny lump of tissue growing out from the placenta of the carpel. At its apex a megaspore mother cell develops and undergoes meiosis to form four haploid megaspores. Normally only one of these cells continues to develop, the rest abort. The single megaspore gets bigger, gaining its nutrition from the nucellus tissue. It undergoes three successive mitotic divisions and forms an embryo sac containing eight haploid nuclei. In the embryo sac the eight nuclei are arranged in a definite pattern. Three cells are situated at the micropyle end of the ovule, making up what is referred to as the egg apparatus, the central cell being the female gamete or egg cell. At the opposite end is another group of three cells called the antipodal cells. The remaining two nuclei, called the polar nuclei, migrate to the centre of the embryo sac.

**Action point**

Draw a pollen sac in TS and LS as seen under the microscope.

**Links**

The process of meiosis is dealt with on pages 142 and 143. Mitosis is dealt with on pages 20 and 21.

**Checkpoint 5**

How many nuclei develop in a pollen grain?

**Action point**

Draw a labelled diagram showing a mature ovule.

**Checkpoint 6**

How many nuclei are produced by meiosis in a single megaspore?

**Exam question**                                   answer: page 154

Describe the male and female parts of the flowering plant, giving a full account of the process of gametogenesis which takes place in them. (15 min)

# Pollination and fertilization

A key feature of the success of flowering plants is their relationship with animals. Pollen grains have no power of independent movement and have to be transferred to the female part of the flower to ensure fertilization. Flowering plants have evolved the strategy of attracting animals, particularly insects, to their flowers, feeding them and exploiting their mobility to transfer pollen from flower to flower. Some plants are pollinated by the action of wind.

## Pollination ●●●

Pollination is necessary in order that the pollen grains, containing the male gametes, are brought into contact with the gynoecium so that fertilization can be achieved. This means that pollen grains must be transferred from the ripe anther to the receptive stigma.

### Self-pollination

In some species self-pollination occurs and the pollen from the anthers of a flower need only be transferred to the stigma of the same flower or another flower on the same plant.

### Cross-pollination

In a large number of species, cross-pollination is more normal, where pollen is transferred from the anthers of one flower to the stigma of another flower on another plant of the same species. There are a variety of mechanisms that ensure cross-pollination or prevent self-pollination, so ensuring outbreeding. Anthers and stigma may mature at different times, they may be at different levels in the flower or there may be separate male and female flowers on different plants.

There are two main methods of pollination:

→ by wind
→ by animal vectors such as insects, birds or mammals

### Wind pollination

Wind-pollinated flowers tend to be inconspicuous, with small petals and sepals, lacking scent or nectaries. Either the stamens are very numerous or there are large anthers present, producing masses of light smooth pollen. The filaments are very long and dangle outside the flower so that the pollen is easily dispersed by air currents. The stigmas are long and feathery, providing a large surface area to trap the pollen. They also often hang outside the flower thereby catching pollen in air currents.

### Insect pollination

Insect-pollinated flowers are large and brightly coloured to attract insects. They have coloured, often shiny, petals and may produce a scent and nectar. The scent attracts the insects and the nectar is collected for food. The stamens are few in number and are fixed inside the flower, where insects must brush past them in order to reach the

**Checkpoint 1**

What is the difference between the definitions of self- and cross-pollination?

**Checkpoint 2**

List three methods of preventing inbreeding in the flowering plant.

**Checkpoint 3**

Why do wind-pollinated flowers produce enormous amounts of pollen?

**Checkpoint 4**

Why do wind-pollinated flowers have feathery stigmas which protrude outside the flower?

nectar. The pollen grains are large, often with elaborately sculptured walls, which help them stick to the bodies of the insects.

## Successful pollination ○○●

Successful pollination results in compatible pollen grains reaching the stigma. If the pollen grains are incompatible, they will not germinate successfully. The stigma produces a sugary solution in which the pollen grains germinate, producing a fine pollen tube. This grows down into the tissues of the style, secreting enzymes which digest some of the cells in order to gain nutrients. The pollen tube nucleus is situated near the top of the tube, with the two male nuclei close behind.

## Fertilization ○○●

The pollen tube grows through the micropyle and passes into the ovule and then to the embryo sac. As the pollen tube penetrates the embryo sac, the tip opens and the two male gamete nuclei enter, the pollen tube nucleus having disintegrated earlier.

→ The first male nucleus fuses with the female, or egg, nucleus to form a zygote.
→ The second male nucleus fuses with the two polar nuclei to form a triploid endosperm nucleus.

These events constitute double fertilization, a unique feature of the flowering plants (see diagram below).

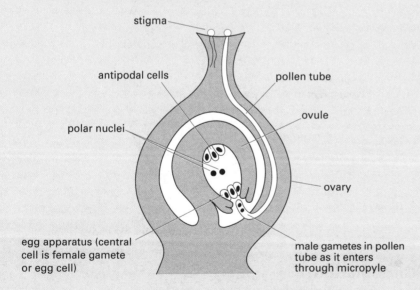

**Test yourself**

Construct a table comparing the features of a wind- and insect-pollinated flower. Include reasons for the various adaptations.

**Action point**

Construct a table of differences between wind- and insect-pollinated flowers.

**Checkpoint 5**

What is the difference between pollination and fertilization?

**Watch out!**

Be careful not to confuse which nuclei fuse together. Draw a diagram of fertilization and using matching colours, colour in the nuclei which fuse together.

**Checkpoint 6**

How is endosperm formed?

**Checkpoint 7**

How many megaspore nuclei are involved in fertilization?

**Checkpoint 8**

What is the importance of the triploid endosperm nucleus?

**Exam question**                    answer: page 154

Describe the way in which fertilization takes place from the time the pollen lands on the stigma. (10 min)

# Seed development and germination

The seed develops from the fertilized ovule and contains an embryonic plant and a food store. After a period of dormancy and when environmental factors are favourable stored food will be mobilized and the seed will germinate.

## Development of the seed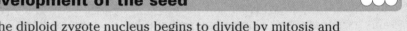

→ The diploid zygote nucleus begins to divide by mitosis and eventually an embryo, consisting of a plumule, a radicle and one or two cotyledons, is formed.
→ The endosperm nucleus also divides mitotically, forming a large number of nuclei and eventually an endosperm tissue develops. These cells become filled with stored food materials, providing reserves for the developing embryo.
→ The integuments fuse together and become the seed coat or testa.
→ The carpel wall, the pericarp, becomes the fruit.
→ The ovule becomes the seed.

## Structure of the seed

The diagram below shows the structure of a broad bean seed.

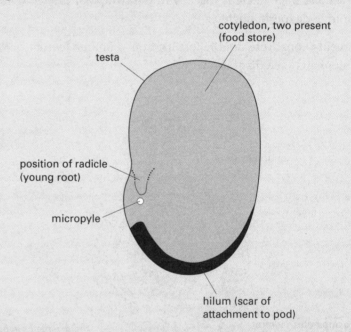

cotyledon, two present (food store)

testa

position of radicle (young root)

micropyle

hilum (scar of attachment to pod)

## Germination

Germination of the seed usually occurs during a period of dormancy. The three main requirements for successful germination are:

→ a suitable *temperature* – this is related to enzyme activity
→ *water* – needed for the mobilization of enzymes, vacuolation of cells and for transport
→ *oxygen* – energy is needed for growth, production of ATP

**Checkpoint 1**

After fertilization what happens to the sepals and stamens of the flower?

**Checkpoint 2**

When the ovary forms, what will
(a) the ovary wall become?
(b) the ovule become?

**Examiner's secrets**

Check your specification for a named seed.

**Checkpoint 3**

Why is oxygen required for seed germination?

## Mobilization of food reserves during germination ●●●

Food reserves in seeds are insoluble in water and cannot be transported in the seedling. They must be broken down into relatively simple soluble substances that dissolve in water so that they can be moved to the growing apices of the plumule and radicle. Water is taken up rapidly by the seed in the initial stages, causing the tissues to swell as well as mobilizing the enzymes. The testa, or seed coat, will rupture as the radicle pushes its way through. The radicle will grow downwards and the plumule upwards.

Secretion of hydrolytic enzymes, e.g. amylase, which hydrolyses starch into maltose, is triggered by the hormone gibberelic acid (GA) produced by the embryo. The GA diffuses to the aleurone layer surrounding the endosperm where protein synthesis occurs (see the diagram of a barley seed below).

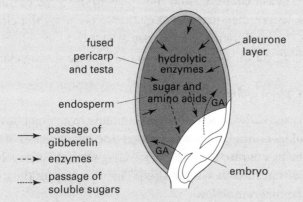

fused pericarp and testa — hydrolytic enzymes — aleurone layer
endosperm — sugar and amino acids / GA
→ passage of gibberelin
---→ enzymes
····→ passage of soluble sugars
GA — embryo

**Checkpoint 4**

Why do seeds need a food store?

**Checkpoint 5**

Why does the seedling establish a root system as the first step in growth?

**Checkpoint 6**

Where is gibberelin produced?

**Checkpoint 7**

Name the layer of the seed where GA triggers protein synthesis.

**Links**

Plant growth regulators are studied on pages 92 and 93.

## The success of the angiosperms ●●●

The angiosperms are the dominant plant group on Earth. The rapid rate at which they achieved diversity may be attributed to several features.

→ Their sexual reproduction is rapid, the interval between flower production and setting of seed being usually a matter of weeks.
→ The protection provided by a closed ovary for the growth of the pollen tube also makes possible an efficient incompatibility system to exclude self-pollination.
→ The fertilization of gametes is independent of water.
→ The unique double fertilization ensures that a food store is only produced if the ovum is fertilized.
→ The cooperative venture between insects and flowering plants has been of mutual advantage.

**Exam question** answer: page 154

Give an account of the formation of seeds after fertilization and describe the germination of seeds. (15 min)

# Human reproduction

In the human reproductive system the gametes are produced in special paired glands called gonads. The male gametes or sperm are produced in the testes and the female gametes or egg cells in the ovaries. You need to study the development of gametes, which is known as gametogenesis.

## Male reproductive system

The male system consists of a pair of testes, which produce spermatozoa; the penis, which is an intromittent organ; genital ducts connecting the two; and various accessory glands, which provide constituents for the semen. Each testis consists of about 1 000 seminiferous tubules, which produce the spermatozoa. The seminiferous tubules also contain interstitial cells, which produce the male hormone testosterone. When sperm have been produced they collect in the vasa efferentia and then pass to the head of the epididymis, where they mature. They then pass along the coiled tube to the base of the epididymis, where they are stored for a short time before passing via the vas deferens to the urethra during ejaculation.

## Female reproductive system

There are two ovaries, each of which contains ova, which are produced in the germinal epithelium. Once formed, the ova develop in follicles. Mature follicles migrate back to the surface when their development is complete so that the ova can be shed. The structure of the ovary is shown in the diagram below.

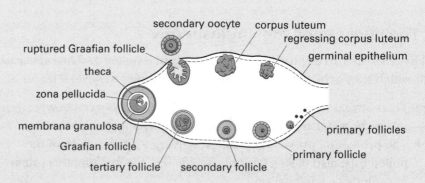

Ova are passed to the fallopian tube or oviduct, which conveys them to the uterus or womb. The uterus is a muscular organ that opens into the vagina.

## Gametogenesis

The cells of the germinal epithelium of both the testis and the ovary undergo a sequence of mitotic and meiotic divisions to form gametes. First there is a multiplication stage which involves repeated mitotic divisions to produce spermatogonia and oogonia. Once formed they grow to full size and then undergo maturation, which involves meiotic division and then differentiation into the mature gametes. The diagram opposite shows the sequence of events in **gametogenesis**.

---

**Action point**

Draw a diagram showing the structure of a testis.

**Action point**

Annotate the diagram of the ovary.

**Watch out!**

Make sure you know at which stage meiosis I and meiosis II take place.

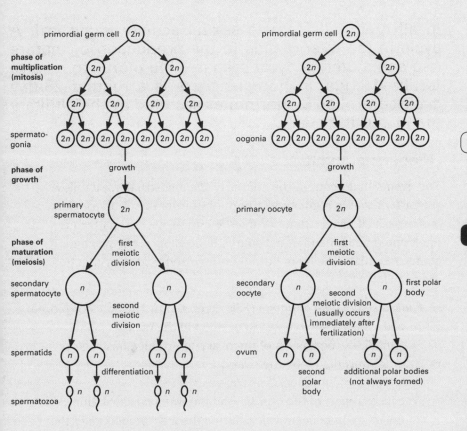

## Formation of spermatozoa

These are produced in the germinal epithelium of the seminiferous tubule and are nourished and protected by the Sertoli cells. They form primary spermatocytes, which then form haploid secondary spermatocytes after the first meiotic division. After the second meiotic division they form spermatids which differentiate into spermatozoa.

## Formation of ova

Oogonia, which are formed before birth, undergo mitosis to form primary oocytes, which then mature just before ovulation. About 2 million of these are formed in the fetal ovary but only about 450 will later develop into secondary oocytes. Before ovulation, a primary oocyte undergoes the first meiotic division to form the haploid secondary oocyte and its first polar body. The secondary oocyte begins the second meiotic division, but this is arrested at metaphase unless fertilization takes place. On fertilization this division is completed to form a large ovum and a second polar body. Once this division has taken place the nucleus of the ovum fuses with that of the sperm to form a zygote, which will then develop into an embryo.

### Links

Details of mitosis are on pages 20 and 21. Details of meiosis are on pages 142 and 143.

### Checkpoint 1

Construct a table to show the differences between the steps in gametogenesis in the male and female.

### Checkpoint 2

What type of cell division takes place in the formation of spermatogonia?

### Checkpoint 3

How many spermatozoa are produced from eight spermatogonia?

### Checkpoint 4

How many ova are produced from eight oogonia?

### Exam question

answer: page 155

Give an illustrated account of the anatomy of the male reproductive system of a mammal indicating the functions of the structures you have described.

(15 min)

# Hormonal control of reproduction

In all mammals female sexual activity is cyclical. A cycle in the ovary leads to the production of mature ova and a uterine cycle prepares the uterus to receive fertilized eggs. In humans there is a regular 28-day cycle controlled by hormones secreted by the pituitary gland and the ovary.

## Menstrual cycle

The events that occur in the course of the menstrual cycle follow a set pattern that is regulated by hormones from the pituitary gland and ovaries. If fertilization occurs, the normal cyclical pattern is interrupted and an additional source of hormones, the placenta, comes into play. The main hormones involved, and their functions, are as follows.

→ *Follicle stimulating hormone* (FSH) secreted by the anterior pituitary:
  → causes Graafian follicles to develop in the ovary
  → stimulates the tissues of the ovary to produce oestrogen
→ *Oestrogen* produced by the ovary:
  → inhibits the production and release of FSH
  → causes repair of the uterine wall following menstruation
  → builds up in concentration during the first two weeks of the menstrual cycle until it stimulates the pituitary gland to produce luteinizing hormone
→ *Luteinizing hormone* (LH) secreted by the anterior pituitary:
  → brings about ovulation
  → causes a Graafian follicle to develop into a corpus luteum which produces oestrogen and progesterone
→ *Progesterone and oestrogen:*
  → inhibit FSH and LH production and therefore stop further follicle development
  → progesterone only causes development of the uterine wall prior to implantation

If fertilization does not take place, the falling levels of FSH and LH allow the corpus luteum to degenerate and the concentrations of progesterone and oestrogen also fall. As a result FSH and LH production is no longer inhibited. The thickened epithelium of the endometrium is shed, producing the menstrual flow.

If fertilization occurs, progesterone and oestrogen, produced initially by the corpus luteum and later by the placenta, are present in high concentration. LH production and then ovulation are inhibited; follicle development and production of oestrogen cease and the uterus lining is maintained.

**Action point**

Draw graphs showing changes in the levels of hormones involved in the menstrual cycle.

**Action point**

Rephrase the list opposite into full paragraphs describing the hormonal control of the menstrual cycle.

**Checkpoint 1**

Explain how progesterone, with smaller amounts of oestrogen, maintains the receptive state of the uterine wall and prevents ovulation.

**Checkpoint 2**

On what day of the cycle does ovulation occur?

## Hormone regulation in the male  ●●●

There is no sexual cycle in the male but gonadotrophic hormones are produced from the anterior lobe of the pituitary gland:

→ *FSH* stimulates sperm development
→ *LH* stimulates the production of testosterone
→ *testosterone* controls the development of male reproductive organs and secondary sexual characteristics

**Checkpoint 3**

Which organ produces testosterone?

### Exam questions
answers: page 155

1  The diagram below shows the relationship between:
   (i)   the pituitary gonadotrophins
   (ii)  the ovarian steroids
   (iii) follicle development
   (iv)  the thickness of the endometrium during the human oestrous cycle

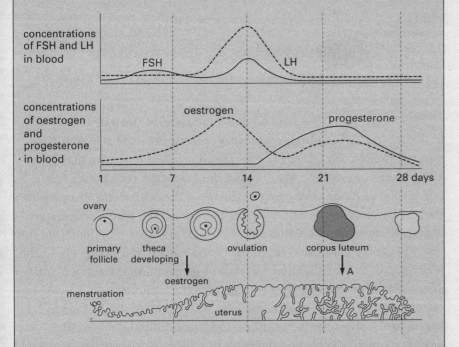

(a) From which cells is the hormone oestrogen secreted and what evidence is there for this in the diagram?
(b) Give *one* effect of the hormone oestrogen on the pituitary gland. Explain your answer with reference to the diagram.
(c) What is represented by the arrow A on the diagram?
(d) What evidence is there from the diagram that the hormone progesterone is involved in a negative feedback mechanism?
(e) How is a knowledge of the effects shown in the diagram involved in the formulation of the contraceptive pill?

(12 min)

2  (a) Name one hormone involved in birth and state its effect.
   (b) What is the role of hormones in lactation?

(6 min)

**Examiner's secrets**

Questions on hormonal control often ask for the analysis of graphs. Many of the answers can be deduced from information provided. Study the graph carefully; always look for peaks and troughs, which mark significant events.
Read the whole question over at least once thoroughly before starting to write your answer.

# Fertilization and development

Internal fertilization ensures that the sperm are deposited in the female's reproductive tract. When a sperm fuses with an egg cell a zygote is formed and this develops by mitosis to form a mass of cells. The zygote is implanted in the uterus wall where it continues to develop. You need to study the placenta and describe its role in the development of the embryo until birth.

## Fertilization ●●●

Sperm are deposited near to the cervix and quickly pass to the top of the fallopian tube by contractions of the uterus and fallopian tube. Sperm are highly fertile for 12–24 hours after release into the female tract but can fertilize an ovum only after a process called capacitation, which takes several hours. It involves changes in the membrane covering the **acrosome**, a thin cap over the nucleus. The structure of the sperm is shown in the diagram below.

**Checkpoint 1**

Why does the sperm have large numbers of mitochondria in the middle piece?

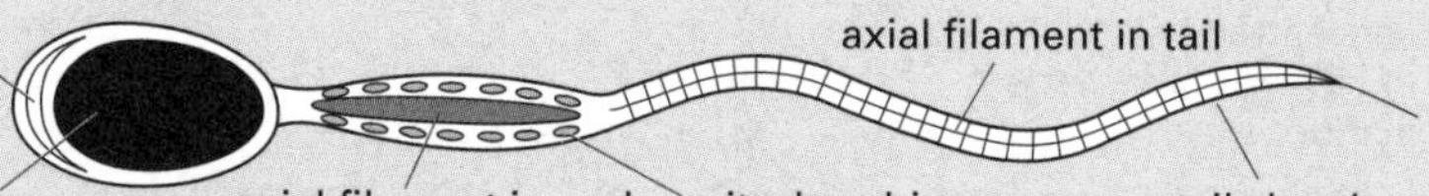

acrosome      axial filament in tail

nucleus    axial filament in neck   mitochondrion    tail sheath

When the sperm reach an oocyte, contact with the jelly coat results in the acrosome membrane rupturing and enzymes are released. The enzymes soften the layers of cells (corona radiata and zona pellucida) surrounding the oocyte. Inversion of the acrosome results in a fine needle-like filament developing at the tip of the sperm and this pierces the already softened portion of the membrane. The whole process is called the **acrosome reaction** and it enables the sperm to penetrate the egg. This entry stimulates reactions of the oocyte that brings about the formation of the **fertilization membrane**, preventing the entry of further sperm. Entry of the sperm also stimulates the completion of the second meiotic division of the oocyte nucleus. The nuclei of the ovum and sperm are drawn together and fuse to form a diploid nucleus. The fertilized ovum is known as the zygote and it begins to divide by mitosis to form the embryo.

**Checkpoint 2**

What is the importance of the fertilization membrane?

## Implantation ●●●

As soon as the ovum is fertilized it begins to divide until a hollow ball of cells, the blastocyst, is produced. The development of the zygote continues during its passage down the fallopian tube, with the outer layer of cells of the blastocyst developing into the embryonic membranes, the chorion and amnion. After reaching the uterus the blastocyst embeds in the endometrium. This is called implantation.

**Examiner's secrets**

You do not need to learn about embryonic development in any detail apart from the role of the amnion.

## The placenta

The placenta develops in the region where implantation occurred. The chorion develops villi, which grow into the surrounding uterine tissue from which they absorb nutrients. These villi form part of the placenta, which is connected to the fetus by the umbilical cord. The amnion develops a membrane around the fetus and encloses the amniotic fluid, which protects the fetus by cushioning it from physical damage.

### Functions of the placenta

The placenta has several important functions.

→ To nourish the developing fetus: in the placenta the maternal blood is brought close to that of the fetus but they do not mix. Finger-like projections of the allanto-chorion grow into the endometrium and become bathed by maternal blood. There is a minimal barrier over a huge surface area, allowing the diffusion and active transport of materials across the placenta. Water, amino acids, glucose, lipids, mineral salts, simple proteins, vitamins, hormones, antibodies and oxygen pass from the mother to the fetus.
→ The removal of wastes: carbon dioxide, urea and other nitrogenous wastes are transferred to the maternal blood to allow their excretion by the mother.
→ To act as a depressurizing unit: it protects the fetal circulation from the high maternal blood pressure.
→ It acts as a filter: the placenta prevents the passage of maternal hormones that could affect fetal development, while allowing some hormones across. It also prevents the passage of some pathogens, although some viruses and bacteria can cross.
→ During pregnancy the placenta progressively takes over the role of hormone production.

**Checkpoint 3**

Why is it important that the blood circulation of the mother and fetus do not mix?

**Examiner's secrets**

You may be required to study the fetal circulation. Check your specification.

**Links**

The menstrual cycle is discussed on page 136.

## Birth

During pregnancy the quantity of different hormones produced changes, with the amount of oestrogen increasing and that of progesterone decreasing. This triggers the birth process and the hormone oxytocin is produced from the posterior lobe of the pituitary gland. This hormone causes contractions of the uterus, which results in the birth of the baby, followed a little later by the placenta, or afterbirth.

**Checkpoint 4**

What inhibits prolactin production?

## Lactation

After the birth the anterior lobe of the pituitary gland produces the hormone prolactin, which causes the mammary glands to produce milk.

---

**Exam question**                                                     answer: page 155

Give an account of the functions of the placenta. (8 min)

# Variation

Variation is the term used to describe the differences in characteristics of members of the same species. Variation may be due to the effect of the environment on an organism but more important, in the long term, are inherited forms of variation that result from genetic changes during sexual reproduction.

## Definition of a gene

In 1866 Gregor Mendel suggested that the characteristics of organisms were determined by 'units' that were handed on from generation to generation. Later these units were identified as genes, which were carried on and transmitted by chromosomes. A gene is the basic unit of inheritance. Genes consist of DNA and have three main characteristics:

→ they can separate and combine
→ they can mutate
→ they code for the production of specific polypeptides

The definition of a gene based on function is 'one gene being the portion of a chromosome that codes for one polypeptide'. Each form of a gene is called an allele. The position of a gene within a molecule of DNA is called the locus. When two identical alleles occur at the same locus on homologous chromosomes, they are said to be **homozygous**; where the two alleles at the same locus differ they are termed **heterozygous**.

It is now known that some genes do not give rise to a product but control the function of other genes.

## Variation

There are two types of variation.

### Discontinuous variation
Discontinuous variation is often controlled by a single gene. This gene may have two or more alleles. There are no intermediate types, e.g. light and dark forms in some moth species.

### Continuous variation
A character within a population which shows a gradation from one extreme to another shows continuous variation, e.g. height. It is influenced by the combined effect of a number of genes (polygenes). The effect of an individual gene may be small but their combined effect is large.

## Origins of variation

### Environmental effects
An organism will inherit genes, giving it a theoretical maximum size, but whether or not this is reached will depend upon nutrition during the growth period and other environmental factors. Thus if organisms of identical genotype are subject to different environmental influences, they show considerable variety. Because these influences are varied they are largely responsible for continuous variation in a population. This is known as non-heritable variation.

**Links**

Chromosome structure is looked at on page 20.

**Checkpoint 1**

Give a general definition of a gene.

**The jargon**

*Homologous* means that in the diploid cell each chromosome has a partner of exactly the same length and with precisely the same genes.

**Checkpoint 2**

Distinguish between continuous and discontinuous variation.

**Checkpoint 3**

Is the presence or absence of ear lobes an example of continuous or discontinuous variation?

Variations due to the effect of the environment have little evolutionary significance as they are not passed from one generation to the next. Much more important to evolution are the inherited forms of variation that result from genetic changes.

## Sexual reproduction

In the long term, if a species is to survive in a constantly changing environment and to colonize new environments, sources of variation are essential. As a result of sexual reproduction variation may be increased when the genotype of one parent is mixed with that of the other. The sexual process has three inbuilt methods of creating variety:

→ The mixing of two different parental genotypes where cross-fertilization occurs.
→ The random distribution of chromosomes during metaphase I of meiosis.
→ The crossing over between homologous chromosomes during prophase I of meiosis.

Although these processes may establish a new combination of alleles in one generation, it is mutations that generate long-lasting variation of a novel kind. However, the occurrence of a useful mutation is a very rare event.

## Mutations

A mutation is a change in the DNA of an organism:

→ it may affect a single gene or a whole chromosome
→ mutations occur randomly
→ most mutations occur in somatic (body) cells
→ only those mutations which occur in the formation of gametes can be inherited
→ mutation rates are normally very small, therefore mutation has less impact on evolution than other sources of variation
→ the rate of mutations occurring can be increased by ionizing radiation and mutagenic chemicals

**The jargon**

*Genotype* means the genetic make-up.

**Checkpoint 4**

Apart from mutations, list three sources of variation.

**Links**

Meiosis is discussed on pages 142–3.

**Examiner's secrets**

An examiner may feel there is scope here for a synoptic question. The question could involve the mechanism of mutation in terms of DNA change and be followed by natural selection. This could lead to extinction and the formation of a new species. See the synoptic question on page 192 in the Resources section.

**Checkpoint 5**

What is a somatic mutation?

**Links**

See also gene and chromosome mutations on pages 148 and 149.

---

**Exam questions**                                      answers: page 156

1   Using the information in the table below, answer the questions:

| Trait | Average difference between pairs of: | | |
|---|---|---|---|
| | Identical twins reared in the **same** family | Non-identical twins reared in the **same** family | Identical twins reared in **different** families |
| Height | 18 mm | 44 mm | 18 mm |
| Weight | 1.86 kg | 4.5 kg | 4.4 kg |
| Head width | 2.85 mm | 4.2 mm | 2.85 mm |
| IQ | 5.9 | 9.9 | 8.2 |

(a) What is the effect of moving an identical twin to a different family? Explain your answer.

(b) Suggest why data were collected for non-identical twins reared together. Explain fully what these data show.

(9 min)

# Meiosis

Meiosis occurs in the formation of gametes. It is referred to as a reduction division because the pairs of chromosomes in diploid organisms are separated in the formation of haploid gametes.

## The process of meiosis

Meiosis is the reduction division that occurs during gamete formation in sexually reproducing organisms. In this division the diploid number of chromosomes ($2n$) is reduced to the haploid ($n$).

Usually there are two cycles of division:

→ firstly, where the chromosome number is reduced
→ secondly, where the two new haploid nuclei divide again in a division identical to that of mitosis

The net result is that four haploid nuclei are formed from the parent nucleus. Like mitosis, meiosis is a gradual process but for convenience it is divided into the four phases, prophase, metaphase, anaphase and telophase, these phases occurring once in each of the two divisions.

### Prophase 1

This stage is similar to prophase in mitosis in that the chromosomes shorten and fatten and become visible but in meiosis they associate in their homologous pairs, and each pair is called a bivalent. At a certain stage in prophase 1 the chromosomes appear double-stranded, as the sister **chromatids** of each chromosome separate and become visible. These chromatids wrap around each other and then partially repel each other but remain joined at certain points called chiasmata. At these points chromatids may break and recombine with a different but equivalent chromatid. This swapping of pieces of chromosomes is called crossing over and is a source of genetic variation (see diagram below).

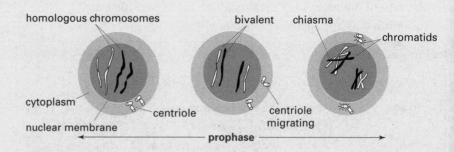

### Metaphase 1

At this stage when the pairs of homologous chromosomes arrange themselves on the equator of the spindle they do so randomly. This random distribution and consequent independent assortment of chromosomes produces new genetic combinations.

## Links

Mitosis is studied on pages 20 and 21.

## Checkpoint 1

Distinguish between the haploid and diploid condition.

## Checkpoint 2

Explain why chromosomes occur in homologous pairs in diploid cells.

## The jargon

A *homologous* pair of chromosomes are of the same length and centromere position possessing genes for the same characteristics at corresponding loci. One of the pair is inherited from the father and the other from the mother.

## Checkpoint 3

What is a chiasma?

## Anaphase 1

Homologous chromosomes are pulled apart: one of each pair is pulled to one pole, its sister chromosome to the opposite pole. The chromosomes reach the opposite poles and the nuclear envelope reforms (see diagram below).

## Telophase 1

The nucleus may enter interphase but in some cells this stage does not occur and the cell passes from anaphase 1 directly into prophase 2. Meiosis II is a typical mitotic division and the result is the formation of four haploid daughter cells.

**Action point**

Construct a table of differences between mitosis and meiosis.

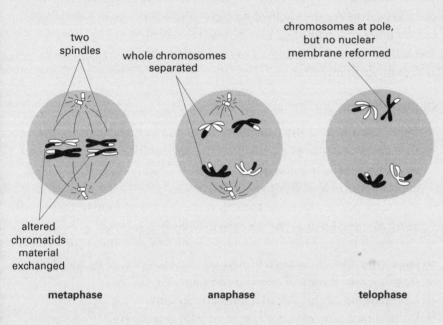

two spindles

whole chromosomes separated

chromosomes at pole, but no nuclear membrane reformed

altered chromatids material exchanged

metaphase      anaphase      telophase

**Checkpoint 4**

Why do gametes formed during meiosis have different chromosome contents?

## Exam questions

answers: page 156

1 The table below describes four events that take place during cell division. Complete the table by ticking the appropriate statement(s) to show which of the events apply to mitosis, meiosis I and meiosis II.

| Event | Mitosis | Meiosis I | Meiosis II |
|---|---|---|---|
| Chromosomes shorten and thicken | | | |
| Crossing over between homologous pairs of chromosomes | | | |
| Double-stranded chromosomes move to the poles | | | |
| Centromeres divide | | | |

(5 min)

2 Write a detailed account of the process of meiosis in plants (the biological significance of the stages within the process is not required). (15 min)

143

# Mendel's laws

Mendel carried out the first documented scientific investigation of inheritance and formulated two laws, which are basic to the science of genetics. You need to understand the principles of monohybrid and dihybrid Mendelian inheritance.

## First law ●●●

Mendel's early experiments were based on selecting pea plants of two varieties that showed clearly separable characteristics, such as tall and dwarf plants and round and wrinkled seeds. From his results he formulated the law of segregation, which states that 'the characteristics of an organism are determined by factors which occur in pairs. Only one member of a pair of factors can be represented in any gamete'. The inheritance of a single pair of contrasted characters is known as monohybrid inheritance.

### Genetic terms

→ *Dominant* allele: of the pair of alleles in a monohybrid cross, the one that always produces an effect on the appearance of the organism when present is the dominant allele. It is usually represented by a capital letter, e.g. T (for tall).
→ *Recessive* allele: the allele that produces an effect only when present as an identical pair is the recessive allele, e.g. t.
→ *Genotype*: the combination of alleles found in an individual.
→ *Phenotype*: the appearance of an organism, determined by the genotype.
→ *Homozygous*: if both alleles are the same, e.g. TT, tt.
→ *Heterozygous*: if the pair of alleles are dissimilar, e.g. Tt.
→ F1: first filial generation; F2, second filial generation.

In monohybrid crosses, two heterozygous individuals will produce a phenotypic ratio of three dominant to one recessive.

## Second law ●●●

Mendel also investigated what happened if he used plants which differed by having two pairs of contrasting characters. In one of his experiments he investigated the inheritance of seed shape (round vs. wrinkled) and seed colour (green vs. yellow). He crossed homozygous plants with round and yellow seeds with homozygous plants with wrinkled and green seeds. When plants grown from these seeds were self-pollinated the seeds produced were of four different types of shape and colour of seed coat. The four types were:

→ round yellow          → round green
→ wrinkled yellow      → wrinkled green

He also found that these types were in the ratio of approximately 9:3:3:1, proportions now known as the dihybrid ratio. This led him to formulate his second law, stating that 'either one of a pair of characteristics may combine with either of another pair'.

The probability of the four alleles appearing in any of the F2 offspring is:

→ round (dominant) $^3/_4$          → yellow (dominant) $^3/_4$
→ wrinkled (recessive) $^1/_4$      → green (recessive) $^1/_4$

**Examiner's secrets**

You should learn the definitions of all the genetic terms.

**Checkpoint 1**

Distinguish between a gene and an allele.

**Watch out!**

Don't confuse genotype and phenotype.

**Checkpoint 2**

What is meant by self-pollination?

**Action point**

Describe the process of a dihybrid cross from the planting of the parent seeds to the gathering of data in the F2 generation.

**Don't forget**

A 'backcross' is a method used in genetics to determine whether a particular dominant characteristic observed in an organism is determined by one or two dominant alleles.

The probability of combinations of alleles appearing in the F2 is as follows:

→ round and yellow = $\frac{3}{4} \times \frac{3}{4} = \frac{9}{16}$
→ round and green = $\frac{3}{4} \times \frac{1}{4} = \frac{3}{16}$
→ wrinkled and yellow = $\frac{1}{4} \times \frac{3}{4} = \frac{3}{16}$
→ wrinkled and green = $\frac{1}{4} \times \frac{1}{4} = \frac{1}{16}$

Below is a genetic diagram to explain the results which Mendel obtained, represented in a Punnett square.

R represents round seed (dominant)    r  represents wrinkled seed (recessive)
Y represents yellow seed (dominant)    y  represents green seed (recessive)

| | | |
|---|---|---|
| parental phenotypes | pure breeding round yellow | pure breeding wrinkled green |
| parental genotypes (2n) gametes (n) | RRYY all RY | rryy all ry |
| F1 genotype (2n) random fertilization self-fertilization | | all RrYy RrYy × RrYy |
| gametes identical from ovules and pollen grains | RY | Ry    rY    Yy |

| ♀ ♂ | RY | Ry | rY | ry |
|---|---|---|---|---|
| RY | RY RY round yellow | Ry RY round yellow | rY RY round yellow | ry RY round yellow |
| Ry | RY Ry round yellow | Ry Ry round green | rY Ry round yellow | ry Ry round green |
| rY | RY rY round yellow | Ry rY round yellow | rY rY wrinkled yellow | ry rY wrinkled yellow |
| ry | RY ry round yellow | Ry ry round green | rY ry wrinkled yellow | ry ry wrinkled green |

F2 genotypes (2n) and phenotypes shown in a Punnett square

9 round yellow    3 wrinkled yellow
3 round green    1 wrinkled green

## Exam questions

answers: page 156

1  Using two named independent contrasting pairs of alleles, show diagrammatically a cross between a homozygous dominant individual and a homozygous recessive individual. With diagrams, show the results of crossing an individual from the F1 generation with:

(a) an individual with an identical genotype

(b) an individual with only recessive genes

Indicate the proportions of the phenotypes in the F2 in each case.

(10 min)

**Watch out!**

Be sure that you understand what is meant by 'probability'. If Mendel's experiment produced 320 seeds in the F2 generation the probability of there being wrinkled and green seeds would be $320 \times 1/16 = 20$.

**Checkpoint 3**

Why was Mendel fortunate in the choice of the pea plant for his experiments?

**Watch out!**

Use the Punnett square to minimize the risk of making mistakes.

**Action point**

Colour in the Punnett square using a different colour for each of the four phenotypes.

**Checkpoint 4**

If Mendel crossed homozygous plants having round and yellow cotyledons with homozygous plants having wrinkled and green cotyledons and he obtained 608 seeds in the F2 generation, how many seeds would have wrinkled and green cotyledons?

**Examiner's secrets**

Once you have practised a number of crosses it is very easy to start missing out stages or explanations and to concentrate on achieving the outcome only. This may make what you have written down in an exam difficult to follow. Even if you achieve the expected outcome, you may not gain full credit. Always carry out genetic crosses in full, giving concise explanations.

# Deviations from Mendel's laws and sex linkage

Not all the characteristics are controlled by single genes which behave independently, as was the case in Mendel's experiments. You need to know about incomplete or co-dominance, and also how sex is determined and how genes on sex chromosomes cause unexpected ratios.

## Co-dominance ●●●

When alleles express themselves equally in the phenotype this is known as co-dominance. In most cases the heterozygote shows a phenotype intermediate between those of the two homozygotes. Examples of co-dominance are:

➜ antirrhinums – red (RR), pink (Rr) and white (rr) flowers
➜ shorthorn cattle – red (RR), roan (Rr) and white (rr) coat colour

The genetic diagram for these crosses is the same as that illustrating Mendel's first law but in the F1 all individuals have the intermediate phenotype.

## Multiple alleles ●●●

Many characteristics in organisms are controlled not by pairs of alleles but by three or more alleles of which any two may occupy the gene loci on the homologous chromosomes. An example of sets of phenotypes controlled by multiple alleles is the ABO blood groups in humans.

## Epistasis ●●●

Epistasis is a form of gene interaction in which a dominant allele at one locus modifies or masks the expression of an allele at a second locus. This can give rise to unusual ratios in genetic crosses. Examples of epistasis are inheritance of certain coat colours in mice and banding in the shell of the snail *Cepaea nemoralis*.

## Sex determination ●●●

Most sexually reproducing animals show two morphologically distinct types, male and female, which are associated with the chromosomes found in the two types. One pair, known as the sex chromosomes, is similar in one sex and dissimilar in the other. The non-sex chromosomes are known as autosomes. In humans the male has dissimilar chromosomes, called X and Y, while the female has two similar X chromosomes.

**Examiner's secrets**

You are only required to study co-dominance in monohybrid crosses.

**Watch out!**

Co-dominance may be expressed using the same letter throughout, e.g. Rr, or by using different letters to represent each allele, e.g. R (red), W (white), RW (pink) in the case of antirrhinum.

**Checkpoint 1**

If roan bulls and cows are mated together what genotypes and phenotypes would you expect in the offspring?

**Checkpoint 2**

Draw a genetic diagram using XX female and XY male as the parental genotypes. What is the expected ratio in the offspring?

## Sex-linked inheritance

The Y chromosome is much smaller than the X and carries very few genes. Therefore in the male any recessive genes carried on the X chromosome will express themselves in the phenotype. This is because they are unpaired and so there is no dominant gene present. This special form of inheritance is known as sex-linkage, an important feature of which is that the male cannot hand on the gene to his sons as they must receive the Y chromosome to become male. On the other hand, all his daughters must receive the recessive gene from him. Females who are heterozygous for sex-linked recessive traits are known as carriers and have a 50% chance of handing on the recessive gene to their sons. To obtain an affected (homozygous recessive) female, the father must be affected and the mother either affected or a carrier. Examples of sex-linked traits in humans are red/green colour-blindness and haemophilia. The process of sex-linked inheritance is shown in the diagram below.

**Examiner's secrets**

You are not required to study autosomal linkage.

**Watch out!**

Don't make the mistake of thinking that sex-linked characters only affect the male.

**Checkpoint 3**

What are the genotypes of the parents who produce a colour-blind female?

| parental phenotypes | normal female | × colour-blind male | | carrier female | × normal male | | |
|---|---|---|---|---|---|---|---|
| parental genotypes (2n) | XX | × | $X^cY$ | $X^cX$ | × | | XY |
| gametes (n) | X | X $X^c$ | Y | $X^c$ | X X | | Y |
| F1 genotypes (2n) | $XX^c$ | $XX^c$ XY | $X^cY$ | $X^cX$ | XX $X^cY$ | | XY |
| F1 phenotype | carrier females | normal male | colour-blind male | carrier female | normal female | colour-blind male | normal male |

| parental phenotypes | carrier female | × colour-blind male | | colour-blind female | × normal male | | |
|---|---|---|---|---|---|---|---|
| parental genotypes (2n) | $X^cX$ | × | $X^cY$ | $X^cX^c$ | × | | XY |
| gametes (n) | $X^c$ | X $X^c$ | Y | $X^c$ | $X^c$ X | | Y |
| F2 genotypes (2n) | $X^cX^c$ | $XX^c$ $X^cY$ | XY | $X^cX$ | $X^cX$ $X^cY$ | | $X^cY$ |
| F2 phenotype | colour-blind female | carrier female | colour-blind male | normal male | carrier females | | colour-blind males |

---

### Exam questions

answers: page 157

1 (a) State two reasons why female haemophiliacs are rarely found in a population.

   (b) Using suitable symbols complete the genetic diagram to show the way in which haemophilia is inherited in the family shown.

| Parental phenotype | Carrier mother | Haemophiliac father |
|---|---|---|
| Parental genotype | ........................ | ........................ |
| Gametes | | |
| Possible offspring genotypes | | |
| Offspring phenotypes | | |

(10 min)

# Mutations

A mutation is a change in the genetic material and is one of the most important topics in genetics. You need to study point (gene) mutations and chromosome mutations and their effect as illustrated by sickle-cell anaemia and Down's syndrome. You also need to appreciate that not all mutations are harmful as exemplified by polyploidy and its importance in crop plants.

**The jargon**

A *mutation* is a change in the amount, arrangement or structure in the DNA of an organism.

## Mutations

Mutations can happen in two ways:

→ DNA is not copied properly before cell division. Sometimes mistakes are made in the copying process so that new chromosomes are faulty. Usually they are small mistakes, involving only one gene, so they are called **gene mutations** or point mutations. However, this can be a serious problem for the individual if a very important gene is affected.

→ Chromosomes are damaged and break. If chromosomes break, they will normally repair themselves (the DNA will rejoin), but they may not repair themselves correctly. This can lead to large changes in the structure of the DNA, and may affect a large number of genes. These are called **chromosome mutations**.

### Gene mutations

Any gene can mutate but rates vary from one gene to another within an organism. Gene mutations are changes in the base pairs within the genes. They can take the form of duplication, insertion, deletion, inversion or substitution of bases. Whatever the change, the result is the formation of a modified polypeptide. A gene mutation in the gene producing haemoglobin results in a defect called **sickle-cell anaemia**. The replacement of just one base in the DNA molecule results in the wrong amino acid being incorporated into two of the polypeptide chains which make up the haemoglobin molecule. The abnormal haemoglobin causes red blood cells to become sickle-shaped, resulting in anaemia and possible death. Haemoglobin S is produced instead of normal haemoglobin by a single base change that causes valine to be substituted for glutamic acid at the sixth position in the β globin chain. DNA codes for glutamic acid are CTT or CTC. Two of the codes for valine are CAT and CAC. In either case the substitution of A for T as the second base would bring about the formation of haemoglobin S. The mutant gene is co-dominant. In the homozygous state the individual suffers the disease but in the heterozygous state the individual has 30–40% sickle cells while the rest are normal. This is called sickle-cell trait.

### Chromosome mutations

These may affect either the number or the structure of chromosomes.

### *Changes in numbers*

This is usually the result of errors occurring during meiosis.

**Checkpoint 1**

Suggest why bacteria are often used in mutation experiments.

**Checkpoint 2**

Explain why gene mutations rarely show up in the phenotype.

**Checkpoint 3**

What are the chances of two parents with sickle-cell trait having a child with sickle-cell anaemia.

**Checkpoint 4**

What is the connection between sickle-cell trait and malaria?

**Don't forget**

Mutations happen naturally but the mutation rate is increased if organisms are exposed to mutagens. It is incorrect to state that mutagens cause cancer. Increased exposure to mutagens increases the **rate** of mutations occurring.

*Aneuploidy* is the loss or gain of a single chromosome. It results in the formation of gametes with either one too many or one too few chromosomes. One pair of chromosomes has failed to separate at anaphase and so the pair has moved to one pole of the cell, the phenomenon of non-disjunction. After fertilization a zygote formed with an extra chromosome is said to be trisomic, e.g. Down's syndrome where chromosome number 21 is trisomic. The error occurs in the production of ova rather than sperm and the incidence of the mutation is related to the age of the mother.

*Polyploidy* involves changes in whole sets of chromosomes. The phenomenon is much more common in plants than in animals. The extra sets of chromosomes interfere with gamete formation, and therefore polyploids often have low fertility, but since many plants can reproduce asexually, the lowered fertility is less important. Polyploidy often gives increased size, vigour and disease resistance, with about half the known species of angiosperms showing evidence of an origin involving polyploidy.

There are two different types of polyploidy.

→ *Autopolyploidy*: there is an increase in the number of chromosome sets within the same species, e.g. if nuclear division takes place but the cytoplasm fails to cleave, a tetraploid cell results. This can be induced artificially by using drugs such as colchicine, which prevents spindle formation. Many important crop plants are autopolyploids, e.g. bananas and sugar beet.
→ *Allopolyploidy*: sometimes hybrids can be formed by combining sets of chromosomes from species with different chromosome numbers. The hybrids are sterile because the total number of chromosomes does not allow full homologous pairing to take place. However if the hybrid has a chromosome number that is a multiple of the original, a new fertile species is formed, e.g. modern wheat.

### Changes in structure

During chiasmata formation, mistakes arise when chromatids break at these points and rejoin with the corresponding piece of chromatid on its homologous partner. There are four types of errors:

→ *deletion* – a small piece of chromosome is lost
→ *inversion* – a piece may be reversed before joining
→ *translocation* – breaks occur in two non-homologous chromosomes and parts of the two chromosomes are exchanged
→ *duplication* – a portion of the chromosome is duplicated resulting in a repetition of gene sequence

These chromosome mutations may cause phenotypic effects and often cause reduced fertility because they interfere with the process of meiosis.

**Test yourself**

Draw diagrams to show structural changes in chromosome mutation.

**Checkpoint 5**

Why can't somatic mutations be transmitted to the next generation?

**Checkpoint 6**

A hybrid may be sterile in the diploid state but fertile when tetraploid. Explain.

---

**Exam question**                                    answer: page 157

1   Modern bred wheat is thought to have evolved from *Triticum durum* (2*n* = 28) which developed from sterile hybrid Z formed from a cross between *Triticum monococcum* (2*n* = 14) and *Aegilops speltoides* (2*n* = 14) a wild grass. Explain fully why hybrid Z is sterile whereas *Triticum durum* is fertile. (10 min)

# Genetic counselling and gene therapy

The techniques developed in genetic engineering involving the manipulation of DNA are also being used in the clinical field with research into changing human DNA (gene therapy), chromosome sequencing, and in the field of forensic science (DNA fingerprinting). Although there are obvious benefits to these approaches in the treatment of genetic disorders, you should also appreciate the possible hazards.

## Genetic counselling

If a family has a history of a genetic defect, unaffected members can consult a genetic counsellor for advice on the risk of bearing an affected child. Advice may be based on:

→ the history of the disorder in the family
→ whether the parents are closely related
→ the frequency of the faulty gene in the general population

### Genetic screening

Once established that there is a risk of passing on a defective gene, there are means of investigating whether a child is affected before it is born:

→ blood tests, e.g. with cystic fibrosis
→ amniocentesis, i.e. removing cells from the amniotic fluid
→ chorionic villus sampling, i.e. early in pregnancy (within 8–10 weeks) tiny samples of fetal tissue are withdrawn from the uterus and cells are cultured and examined under the microscope

On the basis of these tests the parents can decide whether to have the pregnancy terminated.

## Gene therapy

The aim of gene therapy is to treat a genetic disease by replacing defective genes in the patient's body with copies of the healthy gene. There are two possible ways of doing this:

→ *germ-line therapy*, whereby the gene is replaced in the egg
→ *somatic cell therapy*, which targets cells in the affected tissues

### Cystic fibrosis

Cystic fibrosis is due to an autosomal recessive allele. One person in 2000 in Britain suffers from the condition. Cystic fibrosis patients produce thick sticky secretions, which block the pancreatic duct and prevent pancreatic enzymes from reaching the duodenum. There is also a clogging up of the lungs, leading to recurrent infections.

Microbiologists have succeeded in isolating and cloning the gene, which codes for a protein needed for normal functioning. Carriers can be identified using a simple blood test.

It is hoped that an aerosol inhaler could be used to restore the missing gene to the lung cells. This involves wrapping the gene in lipid molecules that can pass through the membranes of lung cells.

### Diagnosis of genetic disorders

This technique is the basis for the diagnosis of many genetic disorders, e.g. sickle-cell anaemia.

→ DNA is isolated from the patient's white blood cells and fragmented using sequence-specific restriction endonucleases.
→ The fragments are separated by **electrophoresis**, then transferred to a filter.
→ The filter is then incubated with a radioactively labelled copy of a specific gene, a gene probe, which combines only with its complementary sequence on the filter.
→ The DNA fragments containing some or all of the probe sequences are then found by using **autoradiography**.

## Genetic fingerprinting

About 90% of the DNA of the human chromosomes has no known function. Individuals acquire different sequences of this non-functional DNA. They vary in length but consist of sequences of bases, 20–40 bases long, often repeated many times. These unique lengths of DNA, known as **hypervariable regions** (HVR), are passed on to the offspring and it is this DNA that is used in DNA 'fingerprinting'. DNA probes have been produced to detect these HVRs at many different loci. The probes developed detect 30–40 different HVRs at once and so the chances of two individuals having the same length is infinitesimally small. The autoradiograph spots reveal a pattern of light and dark bands, which are unique to individuals and is called a genetic fingerprint.

The bands in a fingerprint are inherited from both parents and are used to convict criminals but can also be used in paternity suits. To do this, white blood cells are taken from the mother and the possible father. The bands that belong to the mother are taken away from the bands of the child. If the man is the true father, he must possess all the remaining bands in the child's genetic fingerprint.

**Check the net**

You'll find information on electrophoresis at
www.msn.fullfeed.com/~kendrick/example.htm

**Checkpoint 3**

What is a gene probe?

**Check the net**

You'll find information about the Human Genome Project at
www.ornl.gov/TechResources/Human_Genome/home.html
www-ls.lanl.gov/HGhotlist.html

**Exam question**                                    answer: page 157

What are the principles of human gene therapy? Describe the possible treatment of cystic fibrosis. (8 min)

# Applications of reproduction and genetics

In the past, conventional breeding techniques have been used to improve farm animals, crop and ornamental plants. However, the selection and cross-breeding is laborious, time consuming and sometimes unpredictable. Micropropagation provides a rapid method for obtaining large numbers of genetically identical plants. In the future the quality of farm animals may be improved by laboratory-based breeding techniques involving embryos.

## Micropropagation

Asexual reproduction involves only one organism and the individuals produced are genetically identical, i.e. belong to a clone. The technique of micropropagation is based on the ability of the differentiated plant cell to give rise to all the different cells of the adult plant, i.e. plant cells are **totipotent**. Micropropagation is sometimes referred to as a test-tube plant culture. It is an extremely cost-effective way of producing large numbers of genetically identical plants, which are clones of a single parent. Four steps are generally followed in the micropropagation and are shown in the diagram below.

The advantages of micropropagation are:

➜ large numbers of plants can be grown in sterile controlled conditions ensuring a greater survival rate than would be the case if seeds were planted outside
➜ uniformity of crop
➜ cold storage of large numbers of plants in a small space with reduced heating and lighting costs
➜ perpetuation of unique genotypes
➜ reduced space required for transport
➜ pathogen-free when planted out

### Checkpoint 1

List the differences between asexual and sexual reproduction.

### The jargon

An *explant* is a small segment of stem or bud.
A *callus* is an undifferentiated mass of plant tissue.

### Checkpoint 2

By which cell division process does an explant grow into a callus?

### Checkpoint 3

Why is uniformity in a crop plant an advantage?

### Action point

By referring to the diagram opposite, write out a bullet list to describe the steps in micropropagation.

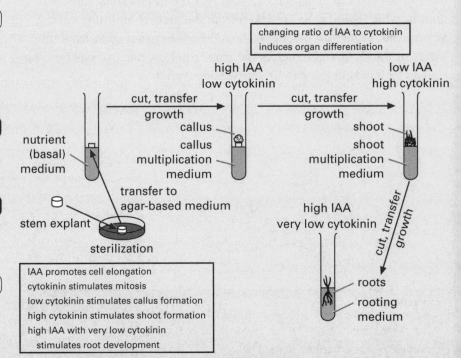

changing ratio of IAA to cytokinin induces organ differentiation

IAA promotes cell elongation
cytokinin stimulates mitosis
low cytokinin stimulates callus formation
high cytokinin stimulates shoot formation
high IAA with very low cytokinin stimulates root development

The disadvantages of micropropagation are as follows.

→ Requires aseptic conditions otherwise bacterial or fungal contamination of the culture medium results with subsequent loss of plants.
→ The plants are genetically unstable, with an increased rate of mutation in medium-grown cells leading to abnormality in the plantlets. Regular inspection is needed to remove any defective individuals, thus labour costs are higher than with traditional propagation methods.

## Cloning of animals

### Nuclear transplants from somatic into egg cells
The cloning of animals may involve the transfer of a nucleus from one individual to an egg from which the nucleus has been removed and the subsequent development of this embryo in a host or surrogate.

### Embryo manipulation
Clones can be developed by taking an embryo at an early stage of development, e.g. after a test-tube fertilization procedure, and splitting the embryo into separate cells. Each of these cells will then develop into a new embryo, genetically identical to the original.

This technique of embryo surgery has enabled farmers to increase their stock. The fertilized egg is bisected at the 2-celled stage and each half is transplanted into different regions of the uterus, allowing twin births to be the rule, not the exception. This technique is also used to conserve rare breeds, where young embryos of young animals are bisected and successfully transplanted into a surrogate mother of a common breed to produce a new individual of the rare type.

### Embryo cloning
This technique allows many genetically identical copies of an animal to be produced. If a mutation occurred in a cow making her a high milk producer, significantly better than other members of the herd, cross-breeding her with a bull would reshuffle her genes with a consequent loss of her unique characteristic. Cloning is the only technique that will conserve her unique features for future generations. The diagram below shows the technique in a mouse.

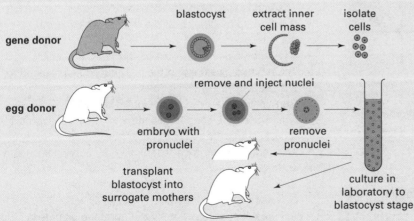

**Links**

The related topic of conservation of rare breeds is discussed on pages 172 and 173.

**Action point**

Discuss the ethical implications of genetic engineering.

**Exam question**                          answer: pages 157–8

Define a clone and describe the principles involved in the cloning of plants and animals. Outline the advantages and disadvantages of this practice. (15 min)

**Action point**

Describe the advantages and disadvantages of cloning animals.

# Answers
## Reproduction and genetics

## Flower structure and gamete development

### Checkpoints

1 Flowers brightly coloured, scented, nectar.
2 Sugar.
3 Large, with projections, and sticky.
4 Nutritive.
5 Three.
6 Eight.

### Exam question

Your answer should include 10 of the following points in a logical sequence.

- Anther produces pollen cells in four pollen sacs.
- Pollen sacs contain diploid spore mother cells each of which divides by meiosis to produce four pollen cells.
- Each pollen cell develops a thick outer wall or exine.
- Mitosis gives rise to a pollen tube nucleus and a generative nucleus.
- The carpel consists of stigma, style and ovary.
- The ovary contains an ovule(s) attached to the wall by a stalk (chalaza).
- The nucellus is surrounded by two integuments.
- The opening between the integuments is called the micropyle.
- The diploid ovule divides by meiosis to form four megaspores, three of which disintegrate.
- The fourth divides by mitosis to produce eight nuclei of the embryo sac.
- There are two polar nuclei in the centre of the cell.
- Three nuclei in antipodal cells at one end of the embryo sac and two synergid cells plus female gamete at the micropylar end.

### Examiner's secrets

This question is in two parts so be careful to devote roughly equal amounts of time to describing the reproductive parts of the flower and to describing gametogenesis. Do not go into details of pollination and fertilization – the question does not ask for these.

## Pollination and fertilization

### Checkpoints

1 Self-pollination is the transfer of pollen to the stigma of the same flower whereas cross-pollination is the transfer to another flower on another plant of the same species.
2 Protandry – stamens ripen before stigma; monoecious – separate male and female flowers; incompatibility; heterostyli; etc.
3 To increase the chance of pollination and fertilization.
4 It gives a large surface area to filter pollen from the air and also for the easy release of pollen into the air.

5 Pollination is the transfer of pollen to the stigma of a flower whereas fertilization is the fusion of gametes.
6 One of the male nuclei from the pollen tube fuses with both polar nuclei in the embryo sac to form a triploid nucleus. This divides mitotically.
7 Three.
8 It becomes the food store or endosperm.

### Exam question

- Pollen tube grows down the style.
- This is controlled by the tube nucleus.
- Followed by the generative nucleus.
- This divides mitotically to give two male nuclei.
- The pollen tube enters the embryo sac through the micropyle.
- One male nucleus fertilizes the female gamete to give the diploid zygote.
- The other male nucleus fuses with the two polar nuclei to form the triploid endosperm nucleus.

## Seed development and germination

### Checkpoints

1 They wither and die.
2 (a) The fruit wall or pericarp.
   (b) The seed.
3 So that energy is made available via respiration for metabolism and growth.
4 To produce substrates for respiration and the building blocks for synthesis.
5 For anchorage and water uptake.
6 Embryo.
7 Aleurone layer.

### Exam question

Your answer should include 10 of the following points in a logical sequence.

- The ovule develops into the seed.
- This forms an outer seed coat or testa.
- This encloses the endosperm and the embryo.
- The endosperm is the food reserve.
- It is formed by division of the triploid endosperm nucleus.
- The embryo develops from the fertilized egg cell.
- The embryo gives rise to the plumule and radicle and to the seed leaves or cotyledons.
- Germination starts with the uptake of water.
- This causes the seed to expand and the coat to rupture.
- It causes enzymes to become active.
- This mobilizes the food reserves in the cotyledons and/or endosperm.
- In cereals the aleurone layer of the endosperm makes the amylase for starch digestion.
- As a result of stimulation by the hormone gibberelin, the growing points (meristems) are reactivated.
- The radicle emerges first followed by the plumule.
- The importance of factors affecting germination.

# Human reproduction

## Checkpoints

1  In the male the germinal epithelium of the seminiferous tubules are active from puberty throughout life; in the phase of multiplication few spermatocytes die; in the phase of maturation and meiosis each primary spermatocyte divides to form two secondary spermatocytes and then four spermatids which mature into sperms. In the female the germinal epithelium of the ovaries is active in the fetus only. In the phase of multiplication most oocytes die. In the phase of maturation and meiosis each primary oocyte divides to form a secondary oocyte and first polar body and further division is delayed until fertilization by a single sperm.

2  Mitosis.

3  32.

4  Eight.

## Exam question

- The male gonads are the sperm-producing testes.
- They are suspended outside the body in the scrotum in order to keep germinal tissue below core body temperature.
- Each testis contains numerous seminiferous tubules.
- These are surrounded by interstitial cells which secrete male hormones.
- The tubules unite to form the vasa efferentia.
- The sperm are stored in the epididymis.
- This leads to the vas deferens or sperm duct.
- This joins the urethra that leads from the bladder through the penis.
- Just before the sperm duct enters the urethra it receives the contents of the seminal vesicles.
- Surrounding the sperm duct/urethra junction is the prostate gland.
- Beyond the prostate the urethra receives the contents of Cowper's gland.
- One function of these glands from the following: secretion of sperm nutrients, activates sperm, secretion of enzymes.

# Hormonal control of reproduction

## Checkpoints

1  After ovulation the concentrations of progesterone and oestrogen gradually increase, and then suddenly decrease. With the fall in level of these two hormones, the endometrium starts to disintegrate and menstruation starts. Oestrogen and progesterone regulate the production of FSH and LH by negative feedback. The high levels of oestrogen and progesterone inhibit the production of the gonadotrophic hormones from the pituitary, and the absence of LH in particular prevents ovulation. Keeping a high level of progesterone helps to maintain the uterus lining.

2  Day 14.

3  Testis.

## Exam questions

1  (a)  Cells of the follicle/theca; the development of the follicle mirrors the rise in oestrogen level.
   (b)  Inhibits FSH production; fall in FSH with rise in oestrogen **or** promotes LH production; positive correlation between oestrogen and LH levels.
   (c)  Maintenance of the lining of the endometrium.
   (d)  While there is a high level of progesterone the pituitary does not produce FSH or LH; when the progesterone level falls the pituitary starts to secrete gonadotrophins again.
   (e)  The pill contains progesterone, which inhibits FSH, preventing the development of ova.

2  (a)  Oxytocin causes the contraction of smooth muscle of the myometrium.
   (b)  Prolactin stimulates the secretion of milk; oxytocin causes the contraction of muscles to force the milk through the nipple.

# Fertilization and development

## Checkpoints

1  To produce ATP to give energy for movement.

2  It prevents the entry of further sperm as only one is required to fertilize the ovum.

3  If the fetus is not the same blood group as the mother it will carry antigens foreign to the mother and she will produce antibodies which will pass to the embryo.

4  High levels of oestrogen and progesterone.

## Exam question

See text.

# Variation

## Checkpoints

1 Genes are the units of heredity that control the characteristics of organisms.
2 In continuous variation there is a gradation of one character from one extreme to another, whereas in discontinuous variation there are a limited number of distinct forms.
3 Discontinuous.
4 Crossing over; independent assortment; chromosomes contributed from each parent.
5 Mutation that occurs in a cell other than a gamete-producing cell.

## Exam questions

1 (a) Increases the range of weight and IQ but not head width and height. Weight and IQ are influenced by both genotype and environment, whereas head width and height are determined only by the genotype.
  (b) As a comparison with identical twins reared in the same family; acts as a control.
     Non-identical twins show a greater range of variation in their characteristics because they have fewer alleles in common.

# Meiosis

## Checkpoints

1 A cell with a full set of chromosomes is diploid; in gamete production during meiosis this number is halved.
2 One member of the pair has come from the male and the other from the female.
3 Points where chromatids remain joined.
4 Random assortment of homologous chromosomes from the mother and father; recombination of segments of chromosomes during crossing over.

## Exam questions

1

| Mitosis | Meiosis I | Meiosis II |
|---|---|---|
| ✓ | ✓ | |
| | ✓ | |
| | ✓ | |
| ✓ | | ✓ |

2 Your answer should include 10 of the following points in a logical order.
  • During prophase 1 chromosomes shorten and thicken and so become visible.
  • The nuclear membrane disappears.
  • Homologous chromosomes form pairs or bivalents.
  • They may entwine forming chiasmata.
  • This is where DNA may break so that genetic material is exchanged, i.e. crossing over.
  • During metaphase 1 each homologous pair line up on the equator of the spindle.

  • During anaphase 1 homologous chromosomes are separated as they move to opposite poles of the cell.
  • This is caused by the contraction of the spindle fibres.
  • The second division takes place with the new spindle forming at right angles to the first.
  • During anaphase 2 the centromeres split as the chromatids move to opposite poles.
  • During telophase 2 the chromatids begin to uncoil, disappear and the nuclear envelope reappears.
  • Four haploid cells are separated as a result of cell plate formation.
  • Bivalents are attached by centromeres.

# Mendel's laws

## Checkpoints

1 A gene controls a character or protein; an allele codes for an alternative form of the same character or protein and occupies the same locus.
2 Pollen is transferred to the stigma of the same flower.
3 The pea exhibits sharply contrasting characters and is relatively unaffected by environmental factors.
4 $608 \times \frac{1}{16} = 38$

## Exam questions

1 You should indicate the correct genotype and phenotype for both parents using an actual example you have studied, e.g. yellow round YYRR × green wrinkled yyrr.
  Indicate the two types of gametes, i.e. YR and yr.
  Indicate the F1 phenotype and genotype, i.e. Yy Rr.
  Cross (a) should be Yy Rr × Yy Rr and should show four types of gametes (use a Punnett square).
  Give the phenotypes and genotypes of the offspring (F2) and the 9:3:3:1 ratio.
  Cross (b), which is in effect the back-cross, is Yy Rr × yy rr.

# Deviations from Mendel's laws and sex linkage

## Checkpoints

1 A ratio of 1 red (RR), 2 roan (Rr), 1 white (rr).
2 1:1.
3 A carrier female and a colour-blind male.

## Exam questions

1 (a) The recessive allele, which is carried on the X chromosome, is sex-linked. It occurs in small numbers in the gene pool, therefore there is little chance of a carrier and a haemophiliac reproducing. Affected females die at puberty or during childbirth.

(b)

| Parental phenotype | Carrier mother | | Haemophiliac father | |
|---|---|---|---|---|
| Parental genotype | $XX^h$ | | $X^hY$ | |
| Gametes | X | $X^h$ | $X^h$ | Y |
| Possible offspring genotypes | $X^hX$ | XY | $X^hX^h$ | $X^hY$ |
| Offspring phenotypes | carrier female | normal male | affected female | affected male |

# Mutations

## Checkpoints

1 All mutations are expressed in prokaryotes but in diploid eukaryotes only the dominant mutations are expressed (due to the heterozygous condition).
2 Because they are recessive.
3 1 in 4, or 25%.
4 In parts of the world where malaria is endemic the condition causes mild anaemia but is an advantage because *Plasmodium* cannot enter abnormal red cells.
5 Because somatic cells are body cells other than the gametes.
6 Chromosomes from different organisms may be unable to pair in the diploid condition, but doubling of the chromosomes permits pairing between homologous chromosomes.

## Exam question

- *T. monococcum* and *A. speltoides* are different species and produce sterile hybrid Z with 14 chromosomes.
- Their chromosomes differ in shape or size and so cannot pair during meiosis; so no gametes form.
- The sterile hybrid can only reproduce asexually.
- When a mitotic spindle fails to form by chance the chromosome number is doubled and a tetraploid form is produced.
- This can now form homologous pairs and can produce gametes or haploid cells during meiosis.

## Examiner's secrets

Plan your answer by rewriting the question in the form of a flow chart. This will set the scene clearly in your mind. You will gain some marks by stating the obvious, e.g. *T. durum* and *T. monococcum* are different species. Suggest why the hybrid Z is sterile. How does it survive and reproduce? By what mechanism does a tetraploid form? Why is it able to reproduce sexually?

# Genetic counselling and gene therapy

## Checkpoints

1 Possible damage to the fetus.
2 1 in 4.
3 An artificially prepared sequence of DNA which can be used as a genetic 'marker'.

## Exam question

The aim of gene therapy is to treat genetic disease by replacing defective genes in the patient's body with copies of the healthy gene. This could be carried out by replacing the gene in the egg, but this approach has not been taken because it could affect the human genome. Instead it may be possible to target somatic cells in the affected tissues.

With cystic fibrosis the approach is to wrap cloned healthy genes in microscopic lipid envelopes and at intervals to spray these on to the surface of the diseased lungs by means of an aerosol spray. It is hoped that the tiny liposome packages will be taken into the cells of the lungs' surface and the genes may enter the chromosomes and be expressed.

## Examiner's secrets

The question does not ask for the symptoms of cystic fibrosis. 'Treatment' could refer to physiotherapy and drugs, but in this context it applies to gene therapy. You should be able to provide a detailed up-to-date answer.

# Applications of reproduction and genetics

## Checkpoints

1 Asexual: no mixing of genetic material, no gametes, one parent, rapid, etc.
  Sexual is opposite of these.
2 Mitosis.
3 Ease of harvesting, all high yielding, etc.

## Exam question

Your answer should include 10 of the following points in a logical order.

- A clone is an asexually reproduced or genetically identical line of cells or organisms.

- A clone may be natural, e.g. bulbs, or may be produced in cell culture/micropropagation of plants.
- Cloning of plants involves the growth of a plant from a part of a plant or from a few cells.
- Cloning of animals involves the transfer of a nucleus from one individual to an egg from which the nucleus has been removed and the subsequent development of this egg.
- The embryo is placed in a host mammal surrogate, e.g. as in 'Dolly the sheep'.
- The embryo can be split before cell differentiation/totipotency (plant/animal) therapeutic cloning of cells for repair and grafting.

The advantages are as follows.
- Cell culture is useful in producing cancer cells for medical research, monoclonal antibodies.
- Produces a single identical genetic line of animal or plant with desirable characteristics.
- Maintains genetic stocks.
- In plants it may be much quicker than seed propagation.

The disadvantages are as follows.
- In mammals and micropropagation in plants the technique is very expensive and can be unreliable.
- In plants disease or the entry of pathogens may cause problems.
- There may be objections on moral grounds.
- There may be inadvertent selection of disadvantageous alleles.
- There may be premature ageing.

**Examiner's secrets**

This is a topical question but do not get carried away with a description of 'Dolly the sheep'. The question is more wide-ranging than this and asks for the principles of cloning in animals *and* plants and you should highlight the differences between them. The question also asks about the advantages and disadvantages so give an equal number of points for and against cloning. There is always a tendency to give details about the morality aspect of genetic engineering. It is a valid point but you must give a balanced answer.

It has been estimated that there are perhaps as many as 10 million kinds of organisms living on the Earth today and an even greater number have become extinct. The huge diversity of life forms has come about due to the process of evolution. It is necessary to study the principles of modern classification and how biodiversity came about. You also need to appreciate the effect that human influences have had on evolution through selective breeding and how humans have contributed to the extinction of species.

## Exam themes

The principles of modern classification
The important features of the five kingdoms
The processes which affect allele frequencies in populations
Examples of how environmental factors can act as stabilizing or evolutionary forces of natural selection
The role of isolating mechanisms in the evolution of a new species
The relative advantages and disadvantages of inbreeding and outbreeding
Selective breeding and an explanation of why it is carried out
The development of resistance in pesticides
The implications of intensive food production on the environment
The importance of genetic diversity
The need to preserve genetic resources

## Topic checklist

| O AS ● A2 | AQA/A | AQA/B | EDEXCEL | OCR | WJEC |
|---|---|---|---|---|---|
| Principles of taxonomy | ● | ● | ● | ● | ● |
| The five-kingdom classification | ● | ● | ● | ● | ● |
| Population genetics | ● | ● | ● | ● | ● |
| Evolution and selection | ● | ● | ● | ● | ● |
| Isolation and speciation | ● | ● | ● | ● | ● |
| Human influences as a selection pressure | ● | ● | ● | ● | ● |
| Biodiversity and causes of extinction | ● | ● | ● | ● | ● |

# Principles of taxonomy

The sorting of living organisms into groups of a manageable size is known as taxonomy or classification. You should know about the principles of modern classification, which shows how organisms may be related through evolution by the number of features they share.

Taxonomy has two main objectives:

→ To classify and arrange species into the broader taxonomic categories.
→ To sort out closely related organisms and assign them to species, describing the characteristics that distinguish one species from another.

**Checkpoint 1**

Write out the scheme of taxa (plural of taxon) within the classification system in order of size, starting with kingdom and ending with species.

**Watch out!**

Homologous should not be confused with analogous characters which are based on similarity in function.

**Watch out!**

The definition of a species has exceptions. The ability to breed is not the sole or essential diagnostic of a species.

**Checkpoint 2**

How can you justify classifying a corgi, an alsation and a labrador as members of the same species?

**Checkpoint 3**

Why is it important for biologists to use Latin names for organisms mentioned in research articles?

**Test yourself**

Write out the scheme of taxa within the classification system in order of size, starting with kingdom.

## Grouping organisms in a hierarchical scheme

The natural classification was devised by Linnaeus in the 18th century. In this scheme organisms are grouped together according to their basic similarities. Such homologous characters are those that have a similar origin, structure and position. A hierarchical system has been devised to distinguish large groups of organisms with a series of rank names to identify the different levels within the hierarchy.

A **taxon** is a level in this classification hierarchy and is an assemblage of organisms sharing some basic features. A species is a group of organisms which share a large number of common characteristics and which can interbreed to produce fertile offspring. Similar species are grouped together into a genus, similar genera into family, families into order, orders into class, and classes into phyla. Phyla are subdivisions of kingdom.

## Binomial system

Organisms are named according to the binomial system introduced by Linnaeus in 1753. Each is named by its genus and species. Latin is used as an international language so that an organism is given precise identification worldwide. Common names, for instance 'tiger', often work well in informal communication, but they can be ambiguous because there are many species of each of the different kinds of organisms. When biologists publish their research, they refer to the organisms they have studied using scientific names to avoid ambiguity, e.g. the tiger belongs to the genus *Panthera* and the species is *tigris*, so its name, written in full, is *Panthera tigris*. The table below describes the classification of the tiger.

| | Taxon | Reasons for inclusion in taxon |
|---|---|---|
| Kingdom | Animalia | heterotrophic, multicellular, no cell walls |
| Phylum | Chordata | notochord, post-anal tail |
| Class | Mammalia | hair, mammary glands, internal fertilization |
| Order | Carnivora | powerful jaws with vertical movement only, large canine teeth, carnassial teeth |

| Family | Felidae | shortened muzzle, fewer teeth than Canidae, retractile claws |
| Genus | *Panthera* | large size (over 3 m long) |
| Species | *tigris* | yellow/tawny fur with dark stripes, larger than lion and more solitary |

**Examiner's secrets**

This is an example only. You are not expected to memorize the taxa or the features.

## Exam questions

answers: page 174

1  The two-spot ladybird, *Adalia bipunctata*, is a common British beetle. Complete the table below that classifies *Adalia bipunctata*.

Kingdom
Phylum                                Arthropoda
                                      Insecta
                                      Coleoptera
                                      Coccinellidae
Genus
Species

(2 min)

**Checkpoint 4**

As one progresses up the hierarchy do the similarities between organisms increase or decrease?

2  Examine the classification flow chart below, then answer the questions.

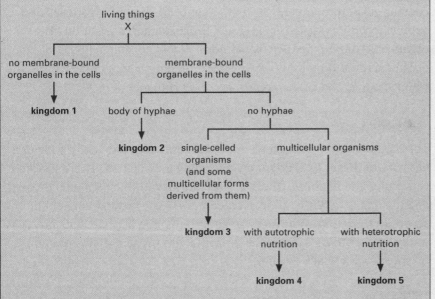

**Don't forget**

Biochemical homology, established by techniques such as amino acid sequence analysis and DNA hybridization studies, provides further evidence for evolutionary affinities within certain groups.

(a) In the five-kingdom classification represented above, why was the first division (X) only correctly established after the electron microscope became available?

(b) The organisms of kingdom 2 have hyphae. What is distinctive about the composition of their walls?

(c) Saprotrophic nutrition is typical of most organisms of kingdom 2. What does this mean?

(d) What are the common structural features of the nuclei of the cells of kingdoms, 2, 3, 4 and 5?

(e) What are the fundamental differences in the nutrition of kingdoms 4 and 5?

(f) Give the names of kingdoms 1–5.

(8 min)

**161**

# The five-kingdom classification

Until recently it was common to include all living organisms into two kingdoms, plants and animals. Because there are some organisms which can be classified in both kingdoms and some which do not fit into either, the five-kingdom classification was proposed. You need to know about the distinguishing features of the five kingdoms.

Living organisms are divided into five kingdoms: Prokaryotae, Protoctista, Fungi, Plantae and Animalia.

## Prokaryotae

The Prokaryotae are unicellular organisms and comprise:

→ bacteria
→ blue-green or cyanobacteria

They have no internal cell membranes, no nuclear membrane, no endoplasmic reticulum, no mitochondria and no Golgi apparatus.

## Protoctista

The members of this kingdom are mostly small unicellular or multi-cellular eukaryotic organisms, with membrane-bound organelles and a nucleus with a nuclear membrane. In this kingdom are found the organisms that are neither plants nor animals nor fungi, in fact all the organisms that do not fit elsewhere! So the kingdom includes algae, the water moulds, the slime moulds and the protozoa.

## Fungi

The members of this kingdom are eukaryotic, the body consisting of a network of threads called hyphae, forming a mycelium. There is a rigid cell wall made of chitin. There are no photosynthetic pigments present and feeding is heterotrophic; all members of the group being either saprophytic or parasitic. In some subgroups, the hyphae have no cross-walls, but in others cross-walls, or septa, are present. Reproduction is by spores which lack flagella.

## Plantae

The members of this kingdom are multicellular and photosynthetic. The cells are eukaryotic, have cellulose walls, vacuoles containing cell sap and chloroplasts containing photosynthetic pigments.

Some of the phyla of the Plantae include:

→ Bryophyta – mosses and liverworts
→ Filicinophyta – ferns
→ Coniferophyta – the conifers
→ Angiospermophyta – the flowering plants

The flowering plants are the predominant plant group. They include all our major food plants. Their flowers have seeds which are enclosed in a

**Watch out!**

Be aware that there are several different approaches to classification, e.g. cladistics and orthodox taxonomy.

**Checkpoint 1**

Why is the Protoctista sometimes referred to as 'the ragbag kingdom'?

**Checkpoint 2**

Some fungi are plant-like in appearance. Why are they now placed in a separate kingdom?

**Checkpoint 3**

Which group of organisms *could* form a sixth kingdom?

fruit formed from the ovary wall. Angiosperms are divided into two classes depending on the number of seed leaves (cotyledons) which they have in their seeds.

→ Monocotyledons – embryo has one seed leaf. The group includes all the grasses, which includes cereals.
→ Dicotyledons – embryo has two seed leaves. The group includes trees such as oak and beech, shrubs and flowers.

## Animalia

The members of this kingdom are multicellular, heterotrophic, eukaryotes, cells lacking a cell wall, exhibit nervous coordination. This group includes the following phyla:

→ Cnidaria (jellyfish)
→ Platyhelminthes (flatworms)
→ Nematoda (roundworms)
→ Annelida (segmented worms)
→ Mollusca (molluscs)
→ Arthropoda (arthropods)
→ Echinoderms (starfish)
→ Chordata

Most of the chordates have a vertebral column and are therefore referred to as vertebrates. The vertebrate chordates are divided into the following classes:

→ fish
→ amphibians
→ reptiles
→ birds
→ mammals

**Examiner's secrets**

You do not need to learn the names of the animal phyla.

**Checkpoint 4**

To which phylum do humans belong?

---

**Exam questions**                                        answers: page 174

1   The diagram below classifies living organisms into five kingdoms.

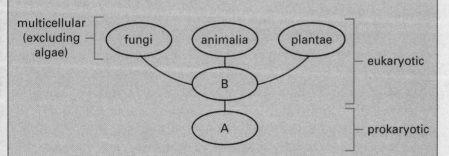

(a) Name kingdoms A and B.
(b) Consider the three kingdoms Fungi, Animalia and Plantae. Give one feature of each that distinguishes it from the other two.

(3 min)

# Population genetics

Population genetics is concerned with the factors that determine the frequencies of alleles in populations of organisms. You will also need to understand the connection between allele frequency and phenotype frequency.

## Gene pool ●●●

A population of organisms reproducing sexually contains a large amount of genetic variation called a **gene pool**. Each organism contains just one of the many possible sets of genes that can be formed from the pool. The gene pool remains stable if the environment is stable. However, if the environment changes some phenotypes will be advantageous and will be selected for, while others will be at a disadvantage and will be selected against. Thus a gene pool is constantly changing, some alleles becoming more frequent and others less frequent. In some circumstances alleles may be totally lost from the gene pool.

**Checkpoint 1**

Define gene pool.

## Genetic changes in populations ●●●

In large populations, assuming there is no selection, the proportion of dominant and recessive alleles of a particular gene remains constant. It is not altered by interbreeding. The Hardy–Weinberg principle is used to calculate allele and genotype frequencies in a population. It can therefore be used to predict the number of defective individuals in a population. The Hardy–Weinberg equation is

$$p^2 + 2pq + q^2 = 1$$

Consider a pair of alleles Aa; let $p$ = frequency of A and $q$ = frequency of a; in the population $p + q = 1$, assuming random mating, then in the next generation AA = $p^2$; Aa = $2pq$; aa = $q^2$.

> *Example*: a recessive allele confers resistance to an insecticide in a particular insect species. Explain how the allele is distributed if 36% of the insect population is resistant.
> rr = 36% or 0.36, i.e. $q^2 = 0.36$, $q = \sqrt{0.36} = 0.6$
> Since $p + q = 1$, $p = 0.4$, $p^2 = 0.16$
> $2pq = 2 \times 0.4 \times 0.6 = 0.48$ or 48%
> Therefore allele distribution is as follows:
> RR ($p^2$) = 0.16 (16%); Rr ($2pq$) = 0.48 (48%); rr ($q^2$) = 0.36 (36%)

**Don't forget**

It is not possible to observe whether organisms are homozygous dominant or heterozygous for a particular characteristic. They all look the same. The Hardy–Weinberg equation allows us to calculate the proportions of the alleles of a particular gene.

Sometimes, variations in gene frequencies in populations occur by chance. This is known as random **genetic drift**. It may be an important evolutionary mechanism in small or isolated populations. Say an allele occurs in 1% of the members of a species. In a large population, say 1 000 000, then 10 000 individuals may be expected to possess the allele. By chance, the population of individuals with the allele will not be significantly altered in the next generation. If, however, the population is much smaller, say 1 000 individuals, only one will carry the allele. By chance, this one may fail to mate and pass on the allele and so it will be lost from the population altogether.

**Checkpoint 2**

Define genetic drift.

**Checkpoint 3**

Suggest factors that can affect allele frequencies over time.

An important case of genetic drift is when a few individuals become isolated from the rest of the species and start a new population, e.g. when a few individuals colonise an isolated island or some new habitat.

These founder members of the new population are a small sample of the population from which they came. By chance, they may have a very different gene frequency. While the founder population remains small it may undergo genetic drift and become even more different from the large parental population. This process is called the **founder effect**. The effect undoubtedly contributed to the evolutionary divergence of Darwin's finches after strays from the South American mainland reached the remote Galapagos Islands.

Disasters such as earthquakes, floods, and fires may reduce the size of a population drastically. The result is that the genetic makeup of the surviving population is unlikely to be representative of the makeup of the original population. By chance, certain alleles will be over-represented among survivors, other alleles will be under-represented and some alleles will be eliminated completely. Genetic drift may continue to affect the population for many generations, until the population is again large enough for sampling errors to be insignificant. Because alleles for at least some loci are likely to be lost from the gene pool, the overall genetic variability in the population is usually reduced.

### Exam questions

answers: page 174

1   Two different varieties of the two-spot ladybird exist, the normal red form and a black form. These two varieties are determined by a single gene with the allele for black body, B, being dominant to that for red body, b. The bar chart below shows the frequency of the allele for black body in autumn and spring samples over a number of years.

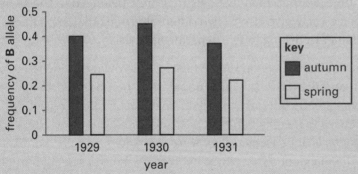

(a) (i)   In the autumn 1929 sample, what was the frequency of the B allele and the b allele?

(ii)   Use the Hardy–Weinberg expression to calculate the expected frequency in the autumn 1929 sample of ladybirds with the genotype Bb and black ladybirds.

(b) Some information about the biology of the two-spot ladybird is given below.

→ The two-spot ladybird spends the winter as an adult. When the weather gets warmer it becomes active and starts to feed and breed.

→ The temperature of the ladybird depends on that of its environment. The higher the temperature, the more active the ladybird.

→ Temperature sensors placed under the wing cases showed that the black areas cool down and heat up faster than the red ones.

Use this information to explain the seasonal changes in frequency of the B allele shown in the bar chart.

(12 min)

### Examiner's secrets

The principle is based on two equations:
$p + q = 1$ (gene pool)
$p^2 + 2pq + q^2 = 1$ (total population).
If you are provided with suitable data always use the $p + q = 1$ equation first to calculate the frequency of alleles. From the figure obtained you can calculate the other parts of the equation.

### Examiner's secrets

You may not be required to use the Hardy–Weinberg principle in calculations; check your specification.

# Evolution and selection

Evolution is the process by which new species are formed from pre-existing ones over a period of time. The basis of contemporary thought surrounding the theory of evolution was first put forward by Alfred Wallace and Charles Darwin. In 1859 Darwin proposed natural selection as the force that causes changes in populations. More recently biologists have realized that natural selection can also maintain variation and therefore stabilize a population.

## Natural selection

Darwin's observations of variation within a population and the tendency for the adult population to be stable in size led to the development of the idea of natural selection. The theory is based on the following observations:

→ variation is found in all populations
→ individuals within a population have the potential to produce large numbers of offspring yet the number of adults tends to stay the same from one generation to the next

From these observations, two deductions were made.

→ There is a struggle for survival with only the 'fittest' surviving. The individuals that survive and reproduce pass on to their offspring the characteristics that enable them to succeed.
→ In time, a group of individuals that once belonged to the same species may give rise to two different groups that are sufficiently distinct to belong to two separate species.

During his voyage on HMS *Beagle*, Darwin found evidence of adaptive radiation in the finch population. He observed 14 species, differing greatly but particularly in beak morphology, reflecting differences in feeding habits. It was suggested that soon after the Galapagos Islands formed they were populated by a flock of finches from the mainland. These then evolved by natural selection to fill the available vacant niches, as other perhaps better-adapted forms that would have been in competition with the finches had failed to reach the Islands.

When environmental conditions change, there is a selection pressure on a species causing it to adapt to the new conditions. This pressure determines the spread of an allele within the gene pool. Darwin proposed natural selection as the force that causes changes within populations. More recently biologists have realized that natural selection can also maintain variation if there is no need to change and therefore stabilize a population. The types of selection are as follows.

→ *Directional selection* is when an environment favours all the individuals carrying one gene at the expense of the individuals carrying an alternative.

**Check the net**

The modern reinterpretation of Darwin's theory is called neo-Darwinism (it is the restatement of the concept of evolution by natural selection in terms of Mendelian and modern genetics) and may be found at
www.panspermia.org/neodazrw.htm
homepages.which.net/~gk.sherman/
baaaaabc.htm

**Checkpoint 1**

Explain why populations remain constant despite the production of large numbers of offspring.

**The jargon**

*Adaptive radiation* is the development from a single ancestral group of a variety of forms adapted to various environments.

**Action point**

Make your own notes on Darwin's finches and how their beaks are adapted to different feeding niches.

**The jargon**

*Selection pressure* is the term given to a factor that has a direct effect on the numbers of individuals in a population of organisms.

→ *Stabilizing selection* results in several different types of individuals being maintained in a population so that genetic variation is maintained.

→ *Disruptive selection* favours both extremes of variants and leads to a divergence of the phenotype, i.e. to the emergence of two distinct forms. This form of selection is rare but is more likely to result in the formation of a new species.

## Polymorphism

Polymorphism is the term used to describe the presence of clear-cut, genetically determined differences between large groups in the same population. Where a population is split into two subpopulations it is the result of conditions that favour more than one phenotype within the population. If gene flow between the subpopulations is prevented, then ultimately each population may give rise to a new species. Examples of polymorphism are colour and banding patterns in the land snail *Cepaea nemoralis* and ABO blood grouping in humans.

**Checkpoint 2**

Is variance increased or reduced by directional selection?

**Action point**

Draw normal distribution graphs to show the different types of selection.

**Checkpoint 3**

How may gene flow between populations be prevented?

**Exam questions**                                        answers: page 174

1  Shell colour and pattern in *Cepaea nemoralis,* the land snail, is genetically determined. Shells may be brown, green or pink. In addition, the whole shell may have up to five dark bands on it. Thrushes eat these snails, which they select by sight.

The table below shows the distribution of some shell types in three different habitats.

| Shell type | Percentage of each shell type present in: | | |
| --- | --- | --- | --- |
| | Grassland (green pasture) | Rough herbage (nettles, long grass and dead stems) | Woodland floor (dead leaves, soil and twigs) |
| Brown | 10 | 23 | 54 |
| Green | 58 | 19 | 12 |
| Pink | 7 | 8 | 4 |
| Banded | 25 | 50 | 30 |

(a) Name the shell type with the greatest survival value on the woodland floor.

(b) Explain the relative abundance in grassland and rough herbage of snails with green shells and pink shells.

(c) Explain the relative abundance of banded snails in grassland, rough herbage and woodland.

(d) In winter the snails hibernate just beneath the soil with the mouth of the shell upwards and covered with white mucus. The thrushes dig out such prey from the soil. Suggest why you would not expect selection to operate on shell colour in these circumstances.

(10 min)

# Isolation and speciation

In theory, any individual in a population is capable of breeding with any other. However, breeding subunits may become separated in some way, i.e. isolated, and evolve along separate lines. If, being reunited after many generations, individuals were found to be incapable of breeding successfully with each other, they would have become a separate species.

## The origins of species ●●●

Within a population of one species there are groups of interbreeding individuals. Each of these breeding subunits is called a **deme**. New species arise when some barrier to reproduction occurs so that the gene pool is divided and different combinations of mutants develop in each pool division. Such a barrier which effectively prevents gene exchange is called an isolating mechanism. If the separation is long term, eventually the two groups will be so different that two new species incapable of interbreeding are formed. For new species to develop from a population, some form of isolating mechanism is required.

There are two main types of isolating mechanisms:

→ *Reproductive isolation within a population* occurs when organisms inhabiting the same area become reproductively isolated into two groups for reasons other than geographical barriers. Species formation occurring in demes in the same geographic area is known as **sympatric speciation**. The barriers to breeding include the following mechanisms:

  → Behavioural isolation – in animals with elaborate courtship behaviour the steps in the display of one subspecies fails to attract the necessary response in a potential partner of another subspecies.

  → Mechanical isolation – the genitalia of the two groups may be incompatible.

  → Gametic isolation – in flowering plants pollination may be prevented because the pollen grains fail to germinate on the stigma whereas in animals sperms may fail to survive in the oviduct of the partner.

  → Hybrid inviability – despite fertilization taking place development of the embryo may not occur.

  → Hybrid sterility (polyploidy) – when individuals of different species breed, the sets of chromosomes from each parent are different. These sets are unable to pair up during meiosis and so the offspring cannot produce gametes.

→ *Geographical isolation between populations* occurs when the population becomes geographically split into separate demes; the evolution of a new species is very probable, given time. This sort of speciation is known as **allopatric speciation**.

Consider the isolation model shown in the diagram at the top of the opposite page. A population of birds with short flight range feed and breed only in the cool conditions of a valley and the lower slopes of two mountains a considerable distance apart.

**Checkpoint 1**

What are the forces for genetic changes in populations that have given rise to new species?

**Examiner's secrets**

You should understand how separation by geographical features, habitat changes, changes in body form and changes in breeding mechanisms may lead to the formation of a new species.

**Checkpoint 2**

Define speciation.

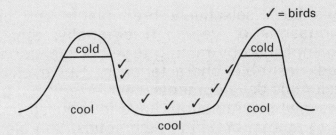

The climate then changes and it gets warmer (see diagram below).

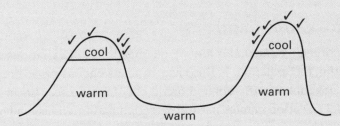

The birds are now confined to the cool mountain tops and the two populations are reproductively isolated. Given a very prolonged period of isolation, genetic drift and directional selection occurs making the two populations genetically distinct.

If the climate reverts to the original situation shown at the top of the page, and the birds are again able to inhabit the valley and lower mountain slopes and the two populations can no longer interbreed, then they have formed two separate species.

Darwin considered that species gradually change over long periods of time from one form to another. It would then be expected that biologists would find intermediate forms between one fossil species and the next in successive rock strata. However, these forms are surprisingly rare and this has led some biologists to believe that new species may arise relatively rapidly (perhaps within a few thousand years) and then remain unchanged for millions of years before changing again.

**Exam question**  answer: page 175

Discuss the part played by geographical isolation in the formation of a new species. (15 min)

# Human influences as a selection pressure

Humans have selectively bred plants and animals for thousands of years. Traditionally, this would be accomplished by mating together individuals that showed desirable characteristics. However, recent new scientific developments, such as embryo transfer and artificial insemination, have introduced improvements to animal breeding techniques. You also need to study how the use of pesticides has applied directional selection to pest populations resulting in the development of pesticide resistance.

## Selective breeding ●●●

Selective breeding of animals and plants makes use of variations which occur within a population. In this case, humans choose organisms showing desirable characters and breed only from these. This process of artificial selection mimics natural selection and provides evidence that selection can lead to the development of characteristics and the production of very distinct forms of organisms, as seen in many domestic animals and plant species. There are two basic methods of artificial selection.

### Inbreeding

Inbreeding occurs when the gametes of close relatives fuse. The problem with inbreeding is that it promotes homozygosity, i.e. it increases the chance of a harmful recessive gene expressing itself, since there is a greater risk of a double recessive individual occurring, e.g. plant species inbred over many generations show a degree of loss of vigour, size and fertility. This is called inbreeding depression. At intervals it is necessary to introduce new genes by outbreeding.

### Outbreeding

Outbreeding occurs by the crossing of unrelated varieties. Outbreeding promotes heterozygosity. It introduces hybrid vigour, where the organisms sometimes grow more strongly. It arises when the new sets of chromosomes are complementary in their effects. Occasionally crosses have occurred between plants of different species, e.g. the development of modern wheat.

There are five main steps in any **breeding programme**:

1. Look for individuals with the characteristics you require, e.g. resistance to disease.
2. Breed together two of these individuals (or self-pollinate if it is a plant).
3. Collect the offspring, and select those that have the characteristics you require.
4. Breed from these offspring.
5. Repeat these steps over many generations.

**Examiner's secrets**

You may be expected to explain *how* the process of selective breeding is carried out in named examples.

**Check the net**

You'll find up-to-date information about artificial selection at www.cd.uiuc

**Checkpoint 1**

What is the main difference between artificial and natural selection?

**Checkpoint 2**

How does outbreeding occur naturally in most flowering plant species?

## Artificial insemination

This involves the collection of semen from male animals which have been selected for desirable characteristics. This is usually diluted and stored by freezing. It is then used to inseminate female animals that may also have been selected for the traits they possess. The main advantages are:

→ it allows more rapid genetic improvement since any farmer can have access to semen from top-quality bulls
→ it eliminates the dangers and costs associated with keeping bulls
→ it provides a wider selection of bulls so inbreeding in a herd can be reduced

The main problem is the need for accurate detection of oestrus to allow insemination at the optimum time.

**Links**

Embryo transfer is discussed on page 153.

## Pesticide resistance

The changes made to natural populations as a result of human alteration of the environment has led to an increase in organisms that do great damage and cause economic losses. Various pesticides have been used to kill pest species. Their use has applied directional selection to the pest populations and this has resulted in the development of resistant populations. Where a single pesticide has been widely and continuously used, many populations now have a high proportion of resistant individuals. Recent control methods have therefore used lower concentrations of a particular pesticide and applications of mixed pesticides. This reduces the selection pressure for a particular resistant gene and resistance evolves more slowly.

**Links**

See also antibiotic resistance on page 123.

**Action point**

Make your own notes on other examples of human influence, e.g. copper tolerance.

### Resistance in rats

Rats, *Rattus norvegicus*, have become resistant to the anticoagulant warfarin which has been used on a large scale to control their populations. A dominant allele (R) at a single locus in rats confers resistance. However, this also confers a requirement for vitamin K.

→ Heterozygotes (Rr) are resistant to warfarin and have only a small requirement for vitamin K.
→ Homozygotes (RR) are resistant to warfarin but have a massive requirement for vitamin K that is difficult to meet.
→ Homozygotes (rr) are killed by warfarin but have a much better chance of survival than RR rats if warfarin is absent from the environment.

This is an example of **heterozygote advantage** where the heterozygotes are favoured by selection and both alleles (Rr) will be maintained in the population, with all three genotypes (RR, Rr and rr) being produced in each generation. Thus in the rat population, both warfarin-sensitive and warfarin-resistant alleles are maintained in areas where warfarin is used as a selective agent.

**Checkpoint 3**

What are the two selection pressures exerted on rats?

**Checkpoint 4**

Why do heterozygote rats tend to survive?

**Checkpoint 5**

Explain, using a genetic diagram, how two heterozygote rats produce all three genotypes in the next generation.

**Exam question**                                   answer: page 175

Explain using examples how human activities can influence evolutionary change. (15 min)

# Biodiversity and causes of extinction

Biodiversity is the variety of species on Earth; extinction is the loss of species. Since the beginning of settled agriculture and the development of urban living, human activities have had various deleterious effects on the environment and on the existence of 'naturally' occurring animals and plants.

**Check the net**

You'll find up-to-date information on biodiversity at www.greenpeace.org

**Action point**

Rephrase these lists into essay style paragraphs.

**Links**

Monoculture is looked at on page 64.

## How farming practice can reduce or enhance biodiversity

The need for increased food production has been satisfied in the following ways:

→ greater use of fertilizers and pesticides
→ increased mechanization
→ increased size of fields
→ improved strains of plant and animal species

The demands of agriculture often conflict with the need for conservation, e.g. farmers remove hedges from their land because:

→ they take up space
→ easier access is needed for large machinery
→ they harbour pests, diseases and weeds

Hedges have considerable conservation value because they:

→ prevent soil erosion by the wind
→ provide a habitat for plants and animals
→ provide food for animals which may not live in the hedgerows

There is therefore a need for the management of farms to ensure sustainability and to reduce the impact on wildlife. This may be achieved by such measures as the control of pesticide use, the maintenance of habitat variety and the prevention of erosion.

## Endangered species

The vast majority of Earth's earlier occupants, including the large and once dominant dinosaurs and tree ferns, have become extinct largely as a result of climatic, geological and biotic changes. At the present time, human activity has taken over as the main cause of species evolution. Many of the larger mammals, e.g. mountain gorillas, giant pandas, tigers and polar bears, are threatened. Their decline in numbers has three main causes: loss of habitat; over-hunting by humans; and competition from introduced species.

Other species are also threatened by additional causes such as deforestation, pollution and drainage of wetlands.

It is now recognized that each species may represent an important human asset, a potential source of food, useful chemicals or disease-resistant genes. There is therefore a need for species conservation, the planned preservation of wildlife. Threatened species are now assigned to one of five categories: endangered, vulnerable, rare, indeterminate or out of danger.

## Conservation of genetic sources

Present-day plants and animals used in agriculture and horticulture have been developed from plants and animals that were originally in the wild. Breeding increases genetic uniformity, with the loss of rarer alleles. In the past breeders may have neglected some important qualities, such as resistance to cold, disease resistance, etc. These need to be added back into highly cultivated varieties, using the wild plants and animals as a gene bank. If habitats and the wildlife they house are threatened, this may no longer be possible. There is also the progressive destruction of the tropical rain forests. Among the many trees and shrubs are some with medicinal properties. The extinction of any plant species before their chemical properties have been investigated could amount to an incalculable loss. In recent years there has been much concern about the loss of gene pools and various legislation has endeavoured to prevent the extinction of endangered species. The following are some of the steps that have been taken:

→ stocks of seeds of 'traditional' varieties of plants are stored in seed banks
→ establishment of sperm banks
→ founding of rare breeds societies to maintain old, less commercial varieties of animals
→ protection and breeding of endangered species in specialized zoos
→ global organizations, such as the World Wide Fund for Nature, mount continuing campaigns to promote public awareness
→ in the UK the Nature Conservancy Council is the government body that promotes nature conservation and gives advice to government and to all those whose activities affect wildlife and their habitats
  → it produces a range of publications
  → it proposes schemes of management for each of the major ecosystem types, endeavouring to conserve species diversity
  → it establishes nature reserves managed by wardens

Education and legislation have also played their part in conservation. Legislation has been introduced to protect endangered species and prevent over-grazing, over-fishing, hunting of game, collection of birds' eggs, picking of wild flowers and collection of plants.

**Examiner's secrets**

When presented with appropriate information, you will need to make balanced judgements between the need to meet the demands for increasing food production and the need to conserve the environment.

**Links**

Conservation is also looked at on page 65.

**The jargon**

*Conservation* is a dynamic process involving management and reclamation.

**Watch out!**

Don't confuse conservation and preservation.

**Examiner's secrets**

You should be prepared to discuss the economic and ethical reasons for conservation.

**Check the net**

You'll find excellent information about conservation issues at www.panda.org

**Exam question**                                    answer: page 176

What are 'endangered species'? For what reasons have some species become endangered? How may species be conserved for future generations? (15 min)

# Answers
## Classification, biodiversity and evolution

### Principles of taxonomy

**Checkpoints**

1 Kingdom, phylum, class, order, family, genus, species.
2 Because they can interbreed to produce fertile offspring.
3 The binomial system uses Latin as an international language so that all biologists are presented with precise identification worldwide.
4 Decrease.

**Exam questions**

1

| Kingdom | Animalia |
|---|---|
| Phylum | Arthropoda |
| Class | Insecta |
| Order | Coleoptera |
| Family | Coccinellidae |
| Genus | *Adalia* |
| Species | *bipunctata* |

2 (a) Since the 1950s microscope studies have been revolutionized by the development of the electron microscope. It enabled the internal structure of cells to be seen in detail and enabled biologists to distinguish between membranous and non-membranous organelles.
(b) Chitin.
(c) Feed on dead organic matter extracellularly.
(d) Eukaryote, membrane surrounding nucleus.
(e) Kingdom 4 make their own food by photosynthesis, kingdom 5 feed on ready-made food.
(f) Prokaryotae, Fungi, Protoctista, Plantae, Animalia.

### The five-kingdom classification

**Checkpoints**

1 It contains all the organisms that cannot be fitted into any of the other kingdoms.
2 They do not possess chlorophyll.
3 Viruses.
4 Chordata.

**Exam questions**

1 (a) A, Prokaryotae; B, Protoctista.
(b) Fungi: lack of chlorophyll, hyphae present, cell walls of chitin.
Animalia: heterotrophic, nervous coordination.
Plantae: autotrophic, possession of photosynthetic pigments, e.g. chlorophyll.

### Population genetics

**Checkpoints**

1 The total of all the alleles of all the genes in a population.
2 Variations in gene frequencies in gene populations that occur by chance.

3 Emigration, reproduction or immigration, genetic drift, mutations.

**Exam questions**

1 (a) (i) B = 0.4 b = 0.6
(ii) Bb = 0.48
Black ladybirds 0.64; you should show that black ladybirds may be homozygous dominant, or heterozygous.
(b) The frequency of black ladybirds is higher in the autumn; black ladybirds absorb heat quicker and become active sooner; they therefore breed more; B allele increases.

### Evolution and selection

**Checkpoints**

1 Predation, competition, etc.
2 Reduced.
3 Geographical and/or reproductive isolation.

**Exam questions**

1 (a) Brown.
(b) Green shells are better camouflaged and are less conspicuous in green pasture than against a mixture of green and brown stems or a foliage of rough herbage. Pink shells are visible and readily selected in both habitats.
(c) The banded pattern makes the shell shape less obvious. Rough herbage presents a mottled background on which banded shells are well camouflaged. Lighter bands show up against the uniform brown of the woodland floor.
(d) The shell is concealed beneath the soil (its colour is hidden), so the thrush comes upon it by chance.

## Isolation and speciation

### Checkpoints

1 Mutations, crossing over during meiosis, independent assortment.
2 The process by which a new species is formed.

### Exam question

Your answer should include all the following points in a logical sequence.

- A description of how the population may become separated.
- For example, when it migrates, due to continental drift, seed dispersal or a geographical feature or physical barrier such as mountain, desert, ocean.
- This separation prevents populations from interbreeding, i.e. prevents the flow of genes.
- The isolated group may possess only a small proportion of the total variation that existed in the original population/gene pool.
- Thus there are different frequencies of alleles.
- Genetic drift may result.
- If environmental selection pressures differ, this will result in differences of phenotype between the two populations.
- This will continue until the original population forms two separate or new species that cannot interbreed if they are brought together.
- Give a suitable example for this, e.g. behavioural, incompatibility, etc.
- Definition of a species – can interbreed to produce fertile offspring.

## Human influences as a selection pressure

### Checkpoints

1 Artificial selection produces marked changes in a short time whereas natural selection changes are gradual.
2 Incompatibility mechanisms, protoandry, etc.
3 Warfarin and vitamin K requirement.
4 They are resistant to warfarin and have a low vitamin K requirement.
5 Rr × Rr using the Punnett square and a simple monohybrid cross.

### Exam question

Your answer should include a detailed description of each of the following topics: use of pesticides and antibiotics and resistance; pollution, e.g. heavy metal tolerance; genetic engineering; artificial selection. You should include some of the following points.

- A resistant gene provides the organism with the ability to avoid or counteract an environmental threat.
- This may be due to attack by a toxin.
- Human activity has introduced toxins into the environment and these act as selective agents.
- Plant example: heavy metal tolerance on spoil heaps.
- Animal example: insecticide resistance or drug resistance in parasites.
- As long as the selective pressure is maintained, resistant alleles provide a competitive advantage.
- The ecological consequences: unprecedented numbers of a pest species which cannot be controlled effectively.
- Pathogen example: antibiotic resistance and its effect on human populations.
- Genetic engineering: some examples of recent genetic modification and its hazards.

# Biodiversity and causes of extinction

## Exam question

Most of the answers to this question can be found in the text. However, there is a need to give detailed examples.

This section is intended to help you develop your study skills for examination success. You will benefit if you try to develop skills from the beginning of your course. Modern AS and A-level exams are not just tests of your recall of textbooks and your notes. Examiners who set and mark the papers are guided by assessment objectives that include skills as well as knowledge.

## Exam board specifications

In order to organize your notes and revision you will need a copy of your exam board's syllabus specification. You can obtain a copy by writing to the board or by downloading the syllabus from the board's website.

AQA (Assessment and Qualifications Alliance)
Publications Department, Stag Hill House, Guildford, Surrey
GU2 5XJ – www.aqa.org.uk

CCEA (Northern Ireland Council for Curriculum, Examinations and Assessment)
Clarendon Dock, 29 Clarendon Road, Belfast, BT1 3BG – www.ccea.org.uk

EDEXCEL
Stewart House, 32 Russell Square, London WC1B 5DN – www.edexcel.org.uk

OCR (Oxford, Cambridge and Royal Society of Arts)
1 Hills Road, Cambridge CB2 1GG – www.ocr.org.uk

WJEC (Welsh Joint Education Committee)
245 Western Avenue, Cardiff CF5 2YX – www.wjec.co.uk

# The AS/A-level specifications

**Action point**

Check with your teacher how your practical skills will be assessed.

**The jargon**

Every A level specification includes synoptic assessment at the end of the A2 course. Synoptic questions draw on the ideas and concepts of earlier units and thus links topics.
Synoptic assessment is dealt with in more detail on pages 188 to 196.

All A-level courses are in two parts, with a number of separate modules or units in each part. Students will start by studying the AS (Advanced Subsidiary). Those wishing to do so will go on to study the second part of the course, called A2. There is some element of choice as to which modules are studied but this may depend on the Awarding Body.

## Advanced Subsidiary

The AS modules are compulsory and they cover the common core of the subject. AS is designed to provide an appropriate assessment of knowledge, understanding and skills expected of candidates who have completed the first half of a full A-level qualification. The level of demand of the AS exam is that expected of candidates half-way through a full A-level course of study, i.e. between GCSE and A-level.

AS may be used in one of two ways:

→ As a final qualification, allowing you to broaden your studies and to defer a decision about specialism.

→ As the first half (50%) of an A-level qualification which must be completed before an A-level award can be made.

Thus the AS specification establishes core principles on which an understanding of biology is based, and at the same time covers relevant topics in sufficient depth to form coherent modules for students who may not continue their study of biology to A-level.

## A-level (AS + A2)

The A-level exam course is in two parts:

→ The AS is the first half of the course and constitutes 50% of the total award. It will normally comprise three teaching and learning modules.

→ The A2 is the second half of the course, the remaining 50% of the total award. It also normally comprises a further three teaching and learning modules which, in addition to introducing new knowledge, extends core principles.

Each teaching and learning module will normally be assessed through an associated assessment unit.

## Assessment units

AS and A2 biology each comprise six assessment units or modules. In both AS and A2 two of the units are assessed by written examination taken at two specific times of the year, January and June. The third component in each case involves a method of practical assessment. Awarding bodies use either centre-assessed coursework or some form of practical examination.

#### What are the main differences between AS and A2?

→ AS and A2 courses are designed so that the level of difficulty steadily increases, with the A2 containing the more demanding concepts.
→ In A2 there is a much greater emphasis on the skills of application and analysis than in AS.
→ A2 includes synoptic assessment.

## Modules and their shelf-life

Any module can be retaken to improve your grade. You may resit any unit *once only* during the time when the results are held in the Awarding Body's unit bank. When you decide to cash in for an AS/A-level award the Awarding Body will use the best result from each unit that you have attempted. Any AS result can be converted into a full A-level award by taking the A2 exam at any time while the specification remains valid.

## Links

While the content of each of the modules at AS is self-contained, during the study of each module opportunities should be taken to enable you to integrate different aspects of the subject. The link feature helps you to achieve this by highlighting connections between different chapters in the revision guide. The specification is designed to allow for the progression from AS to A-level, both in level of difficulty of the topics and in the skill and understanding which are required. Certain topics and concepts which appear as foundation in the AS modules are developed and extended in A2 modules. An example of this is given below:

| AS topics | A2 topics |
| --- | --- |
| molecules . . . | metabolic pathways, e.g. respiration |
| range of organisms . . . | taxonomy |
| DNA/gene technology . . . | applications of gene technology |

## Assessment objectives

For AS and A-level the scheme of assessment will test your ability in the following areas:

→ knowledge and understanding
→ application of knowledge and understanding, analysis, synthesis and evaluation
→ experiment and investigation
→ quality of written communication

At A2 there is an additional category: synthesis of knowledge, understanding and skills. This is the synoptic element, where knowledge and skills from different modules are linked together and assessed.

# Study skills

This spread gives advice on taking notes during your studies and revision techniques to help you prepare for your exams.

## Note taking

Note taking involves condensing speech or writing into an abbreviated form, which nevertheless contains the same *essential* information as the original. This process makes you concentrate on these essentials and in itself is an aid to learning. The second function of note taking is to provide you with an aid to future study and revision. It is much easier to refer back to, or revise from, notes which you have written and organized in your own personal style than from a textbook.

If you are a student attending classes you will probably make most of your notes in the classroom and your teacher will be the main source of information. There are two pitfalls you must try to avoid: do not try to take down every word the teacher says and when making notes from written sources do not copy whole passages. The purpose of note taking is to produce a record of the essential core information. The following are some points of general advice.

→ Use a loose-leaf file so that you can add to, or rearrange, your notes.
→ Space your notes out on the page so that you can add further information later.
→ Pick out key words or headings by underlining or using colour.
→ Always write down book and page references in case you need to check back at a later date.
→ Consider annotated diagrams as an alternative way of presenting information.

### Diagrams

Annotated diagrams are a good way of revising. Starting with a large blank piece of paper can be a good way to focus your mind on a particular topic. You could draw a generalized diagram of a plant in the centre of the paper and then spread more detailed diagrams around it. For example a root hair, a section through the root showing xylem and phloem, an explanation of the cohesion–tension hypothesis of water flow, a section through the stem, a section through the leaf, noting in each case how each organ matches its function. In this way several topics, which may have been studied separately, are brought together. This is an important aspect of revision in the study of biology.

**Action point**

Make your own notes from several sources. Authors vary in the depth of treatment of topics, therefore it is advisable to read a range of publications.

**Examiner's secret**

Teachers recommend using more than one of your senses to aid learning. That's why they often combine 'talk and chalk' or talk with other visual material such as work sheets, TV, etc. Think about using graphic representations of knowledge.

**Checkpoint**

Watching relevant TV is easy and relaxing *but* without note taking or follow-up it does not do much for your communication skills.

## Reading

A-level success depends partly upon the ability to acquire and retain a wide range of knowledge. This is most effectively achieved by reading around the subject; class notes and a textbook alone are inadequate. Try to use more than one textbook as authors vary in their treatment of topics. There are also a number of excellent television programmes on biological topics.

## Language skills

Frequently, students who know a reasonable amount of biology don't do themselves justice in an exam because of their poor exam technique. Much of this is to do with the problem of language. It can make the difference between passing or failing. You need to be able to read, interpret, memorize, understand and convey information concisely in order to answer exam questions. In many of the older textbooks the reader must interpret the complex language used as well as dealing with unfamiliar biological terms. Modern textbooks put much more emphasis on readability, so most of the unfamiliar words presented to the reader will be of a biological nature.

## Understand the basic principles

Many students struggle to remember parts of their subject simply because they do not understand the basic principles involved. Valuable time could be saved by students 'skimming' their notes first to identify such areas in time for them to be dealt with. Facts and relationships are better remembered through a framework of understanding. It is surprising how many students enter exams unaware of what to expect, even to the extent of not having seen a previous year's exam paper beforehand. Every year awarding bodies produce reports on each subject highlighting common misconceptions and providing suggested answers. Your teacher should have a copy of this report. The examiners' comments on candidates' answers are very helpful.

## Revision skills

You must revise regularly. Repetition is an important tool in learning and in preparing for exams. It is not a good idea to leave the revision until just before the exam. Revise each topic as soon as it is completed. Success is achieved through consistent work rather than attempting to commit everything to memory at the last minute.

**Examiner's secret**

Read around the subject. This increases your vocabulary and grammar.

**Action point**

The feature 'check the net' suggests you obtain information using the computer.

**Examiner's secret**

Your teacher should be used effectively for advice and guidance throughout the course. It is essential that your essays and structured questions are set and marked. Learn to regard all criticism as positive and do not repeat errors.

**Action point**

Use the key words. Making your own notes on the words shown in bold in this guide is a valuable exercise. They can be easily referred to and can be used as 'memory joggers'.

**Action point**

Organize your time. Prepare an exam revision timetable. Devote equal time to all topics and leave sufficient time to go over the more difficult topics again.

# Exam papers

At both AS and A-level there is a considerable emphasis on understanding, interpreting and applying knowledge. In addition to the factual content, you need to appreciate the underlying principles of the subject and understand the concepts and ideas associated with them. This, and the information below, will better prepare you for the types of questions you will encounter on exam papers. It will also enable you to understand the principles involved.

## Types of questions

Students are often presented with information on topics which they have not previously encountered and are asked to apply their general biological knowledge in answering questions. The reasons for this are that examiners are required to set questions of different types which test different ways of thinking. These are as follows.

→ Testing knowledge and understanding: the need to know and recall biological facts, name structures, describe their functions, describe how biological processes are carried out and how they are related to each other.

→ Testing skills and processes: here you are expected to apply your biological knowledge and understanding to new situations, e.g. the handling of numerical data in the form of graphs or tables; the analysis and evaluation of numerical data or written biological information; the interpretation of data; the explanation of experimental results.

→ Testing applications of biology: specifications aim to promote an awareness and appreciation of the significance of biology in 'personal, social, environmental, economic and technological contexts'. This type of question not only tests your understanding of the biological principles involved but also your awareness of some of the arguments surrounding an issue, e.g. the use of genetically modified organisms.

A well-balanced exam paper will test a certain proportion of each of the above skills. You must be prepared not only to learn biological facts but also to use your background understanding to interpret and explain unfamiliar biological data, diagrams or descriptions.

## Exam questions

Each module will contain a number of structured questions with one essay question. There may be a choice of essay questions but no more than one out of two per module. Examiners spend a considerable amount of time preparing, revising and refining exam questions. The contents are agreed upon by a committee after each question has undergone thorough and rigorous analysis to minimize the chance of misinterpretation by the student.

**Don't forget**

The Study Guide concentrates on preparing you for the written exam but your experimental and investigative skills will also be assessed in these papers. Revise your practical work in conjunction with your theory!

**Action point**

Do rough calculations of how much time to spend on questions for the papers you are sitting. Use these time allocations when practising answers and develop a confident view of how much you can write in the time.

During revision, when confronted with a mass of information, it is tempting to try to work out which topics are 'going to come up' and leave out certain topics. This is not a good idea with modular exams because the questions are designed to cover every topic. You should be asking yourself what you should do with the knowledge that you have.

## Structured questions

These are the most common types of questions used in AS and A-level biology exams. They can be short, requiring a one-word response, and/or can include the opportunity for extended writing. Remember that the number of lined spaces on the exam question paper, together with the mark allocation at the end of each part question, are there to help you. They are indications of the length of answer expected.

Structured questions are in several parts, usually about a common content. You will experience an increase in the degree of difficulty as you work your way through the question. The first part may be simple recall, perhaps defining a term. The most difficult part is at the end, often asking a question relating to 'technological and social applications of biological principles'.

## Essay questions

All too often students rush into essay questions, writing everything they know about the topic. If candidates took more time, reading the question carefully, they would have a much better chance of finding out exactly what the examiner requires in the answer. When you start each question, plan it out first and write down your plan. This will not only help you to organize your answer logically but will also give you a checklist to which you can refer while writing your answer. In this way you will be less likely to repeat yourself, wander off the subject or miss out important sections.

Look at this sample question:

> Describe the pathways and mechanisms involved in the movement of carbon dioxide in a plant from the atmosphere to the chloroplast in the leaf.

The words to highlight are *describe*, *pathways*, *mechanisms*, *carbon dioxide*, *atmosphere*, *chloroplast*.

The pathway refers to the route taken by the carbon dioxide through stomata, air spaces of the spongy mesophyll layer to the palisade mesophyll cells and into the chloroplasts. The mechanism means 'What makes carbon dioxide take this route?' Carbon dioxide moves in by diffusion because there is more carbon dioxide in the atmosphere than in the chloroplasts during the day since carbon dioxide is being used up in the chloroplasts in the process of photosynthesis.

---

**Examiner's secrets**

Examiners do not set out to trick you. Very often candidates may panic or do not read the question sufficiently carefully, and so 'trick' themselves.

**Watch out!**

Carefully read the general instructions on the exam paper.

**Watch out!**

A2 questions are generally more demanding than those experienced at AS and will require more detailed responses.

**Watch out!**

Papers have time inbuilt for you to read the question carefully. This enables you to decide on which topic the question is focusing and to highlight key words.

**Examiner's secrets**

Where appropriate use well labelled diagrams. Even in 'essay-style' questions this is an excellent way of communicating biology.

# Key words in examinations

Exam questions use various words that key you in on how to approach the question. A list of the words frequently used in examination questions, together with their approximate meanings, is given below.

**Action point**

Prepare a list of biological key words and their meanings. Knowing key words and their meanings gives you a head start. Question openers often ask for such definitions.

**Don't forget**

Always provide units and show your working in calculations. Candidates using calculators often provide the final answer only and lose the marks allocated to the working.

**Examiner's secrets**

*Describe* requires that you give a number of key points. Ideally a separate point for each mark allocated.

**Examiner's secrets**

If the question has been allocated four marks give two advantages and two disadvantages in your answer.

**Annotate**

Give notes of explanation, e.g. each label of a large labelled annotated diagram would include a short description of the function and/or structure as appropriate.

**Brief**

A short statement of only the main points.

**Calculate**

Work out, showing all the stages in the derivation of the answer.

**Compare**

Write about the similarities and differences between two topics, e.g. compare meiosis and mitosis. You need to be organized and without careful planning you will produce a disorganized answer which will be difficult to mark.

**Criticize**

State the faults or shortcomings of, for example, an experiment.

**Define**

State the meaning of, for example, a term without actually using the term itself.

**Describe**

A request for factual detail about a structure or process expressed logically and concisely, e.g. describe the events that occur during the ventilation of the lungs.

**Discuss**

A critical account of the various viewpoints and arguments in the topic set, drawing attention to their relative importance and significance, e.g. 'Discuss the advantages and disadvantages of biological control'.

**Distinguish between**

State the differences between, for example, two or more terms, often for the purpose of identification.

**Explain**

Describe and give reasons for, e.g. explain the concept of autotrophic and heterotrophic nutrition. It is not enough simply to state what these terms mean. You have to explain why some organisms are autotrophs and others are not.

**Graphs**

When a graph is being interpreted it is essential to relate any changes or trends to its biological context, using data as support where possible.

**Illustrate**

Include diagrams or drawings as much as possible.

**List**

A sequence of numbered points one below the other with no explanation.

**Name**

Write the full name.

**Outline**

Give only the main points. This means 'don't go into detail'. But if you have learnt the topic thoroughly you may be tempted to waste time by writing too much!

**State**

A brief concise answer giving no reasons.

**Suggest**

There may be more than one explanation but as long as yours is reasonable it will gain some marks. This means that the question has no fixed answer and a wide range of reasonable responses is acceptable.

**What is meant by**

A definition is usually required. The amount of information to be included is dictated by the mark value.

**Use the data**

Numerical answer required.

## Using past questions

You should make use of past questions, set by the Awarding Body whose exams you are taking, as preparation for the exam. Your school or college may provide you with past papers or you can write to the relevant Awarding Body to purchase copies. Awarding bodies also publish the mark schemes used by examiners. These serve as an excellent guide to what is required on any particular topic.

At the end of each spread in this study guide there are exam questions, and the model answer to each question is given at the end of each section. You should attempt each question *without* reference to your notes, taking notice of the time allocation for the question. After completing the question, mark it yourself. Your wording may not be quite the same but should contain the key words of the answer. If you have not achieved a satisfactory answer, don't be too despondent. Check your notes and rewrite the answer. This way you will improve your understanding and learn from your mistakes. You will probably remember the right answer much better for having gone over it again.

> **Action point**
>
> Practise exam questions. At AS/A-level, knowledge of the facts is not sufficient. You must be able to apply your knowledge to new situations encountered in the exam question.

# Exam planning

One of the keys to exam success is to know how marks can be gained or lost. There are two important aspects to this:

→ ensuring that you follow the instructions on the exam paper
→ understanding how papers are marked by examiners

These will help you to maximize your marks when you tackle the exam paper.

## The examiner's aim

The examiner's job is to produce a fair exam, consistent in style and comparable with equivalent papers in previous years, which tests the biology contained in the specification.

The examiner is looking for certain points in an answer according to an agreed mark scheme. It is important to write legibly and concisely, although the examiner is primarily looking for biological knowledge and application.

**Checkpoint**

Look at the mark and time allocation for each question. This enables you to calculate the importance of each part of the question and assess how much information you need to provide for each part.

**Examiner's secrets**

If you are stuck on one part of one question, highlight that part, move on and come back to it later. You'll be surprised how the answer will then spring to mind.

**Checkpoint**

Plan essay answers – this enables you to organize your thoughts before you start to write.

## Organize your time

Look at the mark allocation for the question. Try to answer every part of every question. If a part question is worth four marks the first two or three points will be easier to access than the fourth but persevere. Time spent planning an essay is time well spent. If there is a choice it also helps to choose the right question. How often has a candidate started an essay question only to find that he/she does not know as much as they thought about that topic? Planning will highlight what you can remember about a topic and whether you have sufficient information to answer all aspects of the question fully.

Allow a little time to read over the question at the end. You will only do yourself justice if you:

→ take your time
→ read through each question carefully before attempting it
→ answer each question to the best of your ability

## Using diagrams

These should always be large, drawn in the correct proportion, fully labelled, have a title and, when appropriate, be referred to in the text of your response. Label lines should be drawn with a ruler so that they do not cross other label lines. They should touch the appropriate structure. Colour and shading should normally be avoided.

There is a difference between a sketch and a diagram. A sketch implies a simple, freehand representation and can be useful to illustrate points relating to experimental situations. Although sketches can be drawn quickly they must be neat; all too frequently they are hurriedly and carelessly drawn. Where the question specifically asks for a diagram, marks will be allocated accordingly in the mark scheme. A diagram must be integrated into, and must make a contribution to, the value of the answer. Detail on the diagram should be limited to the points you wish to make.

→ Don't leave the examiner to work out which part of the diagram is answering the question.
→ Do not rely solely on your labelled diagram.
→ Develop and expand points in your written answer so that the examiner is in no doubt that you realize the relevance of the diagram to your answer.

Using a well-labelled or annotated diagram can be a valuable aid to an exam response. A diagram can often record a considerable amount of information in a short space thus saving several lines of description which in turn saves time. But watch out – don't spend too long on your diagram, particularly if it relates to only part of a question. Try to practise in advance drawing some of the key diagrams you may need.

## Using flow charts

Flow charts are an excellent aid to revision, used to summarize a topic. However, you cannot answer an exam question solely by using a flow chart or a metabolic pathway. You should include some descriptive text to explain the purpose of the chart or pathway, making it clear to the examiner what the chart is meant to show.

# Synoptic assessment

One of the aims of A2 assessment is to enable candidates to show knowledge and understanding of facts, principles and concepts from different areas of biology and to make connections between them. You must understand what is meant by 'synoptic assessment' if you are to be fully prepared for your A2 examinations.

## Synoptic assessment

→ Involves the drawing together of knowledge, understanding and skills learned in different parts of the AS/A2 biology course.
→ May require you to apply knowledge of a number of areas of the course to a particular situation or context.
→ May require you to use knowledge and understanding of principles and contexts in planning experimental work and in the analysis and evaluation of data.
→ Accounts for 20% of your A-level course, but because it takes place at the end, it makes up 40% of the A2 course.
→ Takes place at the end of your course, mainly in the final module.

## Why do A2 exams contain a synoptic element?

→ Modular exams were introduced for a variety of reasons, one of which was to allow students to study a 'theme' in a manageable block of information.
→ However, in modular courses there is a tendency for candidates to gain success in a module then forget the concepts they have learned as soon as the exam has been taken.
→ Synoptic questions require you to transfer concepts from one module to another and to make connections.
→ Having an overview of a subject is thought to be a better preparation for higher education or employment.
→ Being able to link information gives you a greater understanding and is more satisfying and fulfilling, allowing you to contribute more effectively in problem-solving situations.

Consider the analogy of a computer game program. Writing the various programs to create the characters, scenes and situations is an achievement in itself but seeing how the characters interact in the game as a whole is far more satisfying.

## How can I gain synoptic skills?

→ Biology is inherently cyclical and therefore synoptic. During your study of A2 biology you naturally apply many of the ideas and skills learned during AS biology, and even your GCSE course! So in studying A2 biology you will have been using synoptic skills without realizing it.

# What type of questions can I expect?

Different awarding bodies may use different strategies for their synoptic assessment:

→ **Structured** type questions – extended answers based on biological themes drawn from AS and A2 biology. To answer this type of question you use ideas and skills that permeate biology and your task is to interpret the information supplied in the question.
→ **Essay** questions – a long open-ended question, where you make the links and use connections between different areas of biology. Some examples are given below:
  → Why is ATP considered to be 'the energy currency' of the cell?
  → Consider the role of water in plants and animals.
  → Discuss the influence of humans on the environment.
→ A problem set in a **practical** context.

# How can I prepare for synoptic questions?

→ For your specification check out which module will contain synoptic questions.
→ Revise the basics of earlier modules and as you do so list topics that link with those that you have recently studied at A2.
→ Carry out regular revision throughout the course.
→ Make use of past question papers, particularly specimen questions provided by your Awarding Body.
→ As you can expect to encounter new contexts which draw together different ideas, read generally around the subject and 'check the net'.

# Are there any obvious synoptic links?

Yes there are! Certain topics lend themselves more easily to synoptic assessment as given in the following guide:

→ Ecological energy transfer and energy harnessing.
→ Population growth and microbiology.
→ Digestive system and enzyme action.
→ Cell structure, enzymes and energy harnessing.
→ Food chains, pest control and biodiversity.
→ Biochemistry of DNA, mutations, cell division and evolution.

**Examiner's secrets**

With essay questions you make the links between different areas of biology yourself but with structured questions the context has been chosen for you.

**Action point**

Check with your teacher, the specification and look at past papers to determine the type of synoptic questions you can expect.

**Action point**

As you review your notes make your own synoptic links.

# Examples of synoptic questions

## 1. A short structured question

This question uses specific information provided.

The graph below shows the percentage of barn owls found with rat poison in their livers since 1983. The owls are eating resistant rodents (rats and mice) which have eaten the poison, but have not themselves been killed.

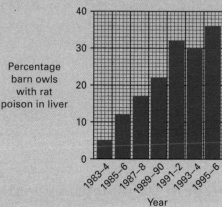

(a) (i) What is the change in percentage of barn owls found with poison in their livers from 1983–4 to 1995–6?

(ii) How have the rodent populations that are resistant to the poison arisen?

(b) Pesticides are used to kill insects that cause damage to crops. The farmers of Warangal, a district of India, planted large areas of arable land with only cotton for a number of years. The crop is prone to pest attack and disease. *Spodoptera litura* is a caterpillar that attacks cotton. Despite frequent applications of pesticide the caterpillar has continued to destroy the crop and is now eating other vegetable crops as well.

(i) Name the taxon level represented by *Spodoptera*.

(ii) What term is applied to the type of crop-growing described?

(iii) Why does growing a crop on this scale often lead to increased attack by pests?

(iv) Suggest two ways that the farmers could have altered their farming practices in order to reduce the attack by pests.

(10 min)

### Answer to question 1

(a) (i) 36 – 5 = 31%

(ii) Variations/mutations in the genes of rodents allow them to survive the poison/a mutation has arisen that enables the rodent to produce an enzyme which breaks down the poison; these advantageous genes are passed to the offspring which are then also resistant.

(b) (i) Genus

(ii) Monoculture

(iii) Pests have no/limited competition for food; so the reproduction rate is fast/exponential; allowing easy access/transmission from plant to plant.

(iv) Plant other crops between the cotton; use biological control/a natural predator; use a number of different pesticides to reduce the chance of the pest becoming resistant to all of them; use integrated pest management. (Any two responses)

## 2 An essay question

This tests your breadth of knowledge and understanding.

> Discuss the statement that 'Living organisms can acquire and transform energy, using it to synthesize complex molecules'.
>
> (20 min)

### Answer to question 2

The following is an outline of the type of answer expected.

Energy from sunlight is captured by the plant using pigments, e.g. chlorophyll, and transformed in the chloroplast to chemical energy, resulting in the production of ATP.

During the light-independent stage of photosynthesis the energy is incorporated into hexose sugar. This may be polymerized into starch. The plant must synthesize all organic materials from the intermediates and products of photosynthesis.

Energy flow through the ecosystem – a brief discussion of trophic levels and food chains, food webs. Energy losses by respiration and excretion. The role of decomposers in the release of energy 'locked' in dead organic material.

Respiration – Krebs cycle and the electron transport chain resulting in the release of energy from glucose in the form of ATP. The use of ATP for active transport, muscle contraction, etc.

**Watch out!**

All essays, whether synoptic or based on the specialized option within a specific module, usually include marks for correct punctuation, spelling and grammar.

**Examiner's secrets**

Energy is involved in so many processes that it is very likely to be included in a synoptic question. Plan the essay first. In this type of essay it is the breadth of knowledge rather than the depth that will gain marks. Don't be tempted to write all you know about respiration and not leave yourself sufficient time to cover the other areas required in the answer.

# 3 A comprehension question

In this type of question a passage from a scientific journal is used as a basis to test your knowledge of the topic by asking you to explain phrases within the text. At the same time it is a synoptic question in that it gives you the opportunity to show your understanding and make connections between the ecological concepts of populations and competition and the concepts of evolution, selection and speciation. These are linked together by the effect of climatic change on the numbers of plant and animal species and how a change in the balance of populations can affect a species over time.

Read the following information, which is based on an account in Steve Jones's book on evolution – *Almost Like a Whale* (Anchor Press, 2000).

> On one of the Galapagos Islands two types of Darwin's finches are found. One is a large, principally ground-dwelling bird, feeding mainly on hard nuts. The other is smaller and has adapted to feeding on soft fruits and berries. The fossil record shows that both of these two types have been in existence for thousands of years.
>
> Every few years the winds of the Pacific Ocean change direction, creating a current of warm tropical waters, called El Niño, which moves eastwards towards America. In the Galapagos Islands El Niño causes a huge increase in rainfall and a corresponding growth in lush and abundant vegetation. Many fertile hybrid finches are hatched which are intermediate between the large and small forms.
>
> As El Niño ebbs away, drought follows the torrential rain and most of the finches die.

(a) Describe the main adaptation that the larger finch would show to enable it to feed successfully.

(b) (i)   The two types of finch are examples of different
        Genera    Species    Varieties    (Circle your choice)
    (ii)  Explain your answer.

(c) (i)   What effect would you expect the arrival of El Niño to have on the finch population?
    (ii)  Explain your answer.

(d) Suggest why 'many hybrids are hatched'.

(e) (i)   Under what conditions is competition most severe
        normal    El Niño    (Circle your answer)
    (ii)  Explain your choice.

(f) (i)   What does the fossil record tell us about what happens when El Niño 'ebbs away'?
    (ii)  Explain why this happens.
    (iii) Name the evolutionary process this illustrates.

(g) Suggest what might happen to the finch population **over a long period of time**.
    (i)   I   If El Niño became permanent.
         II  Explain your answer.
    (ii)  I   If El Niño disappeared permanently.
         II  Explain your answer.

(20 min)

**Answer to question 3**

(a) Strong/more powerful/large (not sharper).

(b) (i) Circle 'varieties'.

    (ii) They can still interbreed; to produce fertile offspring.

(c) (i) The population would increase.

    (ii) There is a large increase in vegetation and so there is more food; the consequence of this increased food supply is that there will be increased breeding success.

(d) Since there is an increase in numbers the two isolated populations are more likely to come in contact with each other.

(e) (i) Circle 'normal'.

    (ii) Under normal drought conditions there is less food available so there is an increase in competition for the reduced food supply.

(f) (i) The intermediate forms do not survive/all the hybrids disappear.

    (ii) The specialized feeders survive because they are best adapted to compete for the limited food supply caused by the drought; the intermediate forms starve. (2 out of the 3 points needed)

    (iii) Natural selection/survival of the fittest.

(g) (i) I The two populations might merge into one large intermediate group.

      II With abundant food the specialized feeders might be at a disadvantage.

    (ii) I Two species would merge, incapable of hybridization.

      II With no periodic abundance, isolation of two forms becomes permanent.

## 4 Drawing a graph from data

The question requires recall, the analysis of results, the conversion of results into graph form, the application of principles.

Enzymes are difficult to recover for reuse at the end of an industrial process. This is because they have to be coagulated before they can be filtered off. However, enzymes are easily reused if they are immobilized. They are trapped and held within a framework of cellulose, which is permeable to the substrate and products of the reaction.

The table below shows the effect of temperature on the maximum rate of reaction (arbitrary units) of an enzyme in its free and in its immobilized state.

| Temperature °C | Free enzyme | Immobilized enzyme |
|---|---|---|
| 0 | 0.5 | 0.5 |
| 10 | 6.0 | 8.0 |
| 30 | 20.0 | 22.0 |
| 40 | 24.0 | 24.0 |
| 50 | 15.0 | 24.0 |
| 70 | 0.0 | 8.0 |
| 80 | 0.0 | 0.0 |

(a) (i)  Plot the data in a suitable form on graph paper.
   (ii)  Explain the difference in reaction rate at 0 °C and 70 °C for the free enzyme.
   (iii)  Describe two differences between the effects of temperature on the immobilized and the free enzyme.
   (iv)  Suggest how trapping and holding an enzyme in a framework of cellulose microfibrils, as shown in the simplified diagram below, can explain the differences you have described in part (iii).

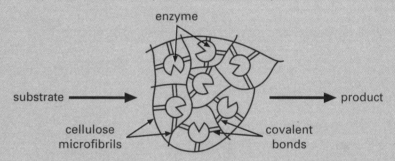

   (v)  Suggest one advantage of immobilizing enzymes other than making them easier to reuse.

(b) Immobilized enzymes can be used in the continuous commercial production of fructose sugar syrup from starch. State three advantages of continuous culture over batch culture.

(c) Suggest why antibiotics can only be prepared by batch culture.

(20 min)

**Answer to question 4**

The following is the type of answer expected.

(a) (i) Graph. The axes should be the correct way round with 'temperature' on the *x*-axis and 'rate of reaction' on the *y*-axis. Labelling must include units. The scale should be correct with the graph filling most of the paper. All plots have to be correct and all lines connecting the points must be neat (not 'sketchy') and must be labelled.

   (ii) There is little kinetic energy at 0 °C with few collisions between the substrate and active site. Reference should be made to activation energy. At 70 °C the shape of the active site is altered/the enzyme is destroyed or denatured.

   (iii) The activity of immobilized enzyme is greater between 0 °C and 40 °C and it has a lower optimum temperature than that of the free enzyme. There is a larger increase in the rate of reaction between 0 °C and 30 °C for the immobilized enzyme than for the free enzyme. The optimum temperature of immobilized enzyme covers a wider range (40–50 °C). Above 40 °C the immobilized enzyme is more active. The free enzyme is inactive at 70 °C when the immobilized is active. Immobilized enzyme is more active at all temperatures except 40 °C.

   (iv) Maintains shape of enzyme/3D structure/stability/enzyme cannot move/denatured.

   (v) Product is not contaminated with enzyme/can be drained away. Enzyme is protected against changes in pH/temperature. Several enzymes with differing pH/temperature optima can be used together/a wider range of conditions can be tolerated.

(b) (Maximum) yield maintained/more product in unit time/constant supply of product over a long period of time (weeks). No need to empty, clean and sterilize culture vessels as often. More substrate/nutrients can be added and products withdrawn without interrupting production. A purer/more uniform product is formed. Smaller vessels can be used.

(c) Antibiotics are produced as growth slows, not when growth is at a maximum. Antibiotics are associated with a particular phase. They are secondary metabolites. Organisms may be unsuitable/unstable for growth by continuous culture.

## 5 Understanding and interpretation of diagrams

### Watch out!

The purpose of bold type in a question is to help you. Pay special attention whenever you come across it.

### Examiner's secrets

This question asks you to redraw a transverse view from a diagram of a longitudinal section so that you can demonstrate your understanding of the three-dimensional nature of the cell as well as making a link with your practical work.

The diagram shows structures that are scattered on the **lower** surface of the leaf and allow the exchange of gases to take place.

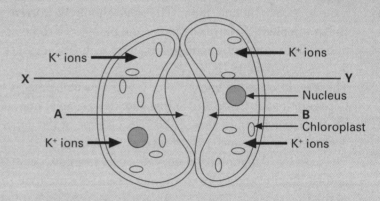

(a) (i) Name the features labelled **A** and **B**.

(ii) Draw a simple diagram of a transverse section along **X–Y** to show the **relationship** between this structure and the two cell types that lie closest to it.

Label **each** of the **cell types** shown in your diagram.

(b) One theory suggests that the space labelled A becomes larger as a result of the influx of K$^+$ ions into the cells shown in the diagram.

(i) Name the **process** that results in the movement of K$^+$ ions.

(ii) Describe how this influx of K$^+$ ions results in an increase in the size of **A**.

(iii) **Circle** the line that describes the stage that provides a source of energy for this process.

| | |
|---|---|
| Calvin cycle | Cyclic photophosphorylation |
| Light-independent reaction | Photolysis |

(10 min)

### Answer to question 5

(a) (i) **A** Stoma/stomatal pore or aperture. **B** Cell wall.

(ii) Diagram showing TS lower surface of leaf with labelled cells; two guard cells with epidermal cell on each side and two or three spongy mesophyll cells.

(b) (i) Active transport.

(ii) (The influx of K$^+$ ions results in) The water potential of the guard cells becoming more negative; water enters; by osmosis; the guard cell becomes turgid/increases in volume.

(iii) Circle 'cyclic photophosphorylation'

# Glossary

**Action point**

This is a selection of biological words you will encounter and is not a complete list of all the words you are required to know. As you work through each chapter, list the key biological words you come across and write out their definitions.

**absorption spectrum**

The range of a pigment's ability to absorb various wavelengths of light.

**acetyl CoA**

The entry compound for the Krebs cycle in cellular respiration.

**action potential**

A rapid change in the membrane potential of a nerve cell, caused by stimulus-triggered, selective opening and closing of sodium and potassium ion channels.

**active site**

The specific portion of an enzyme that attaches to the substrate by means of weak chemical bonds.

**active transport**

Process by which substances are moved across a cell membrane against a concentration gradient, with the help of energy input and specific transport proteins.

**adaptation**

A feature that increases the chance of survival of an organism in its environment.

**adaptive radiation**

The emergence of numerous species from a common ancestor introduced into an environment.

**allele**

An alternative form of a gene.

**Examiner's secrets**

Exam questions often ask you to define a term. This is the examiner's way of focusing your mind on a topic. It's also an easy way to pick up marks!

**Examiner's secrets**

When students first study AS/A2 biology many are surprised that there are so many terms to learn. Expect to define these terms to gain marks in both structured and essay questions in exams.

**antibiotic**

A chemical that kills bacteria or inhibits their growth.

**antibody**

An antigen-binding immunoglobin, produced by β lymphocytes, that attacks non-self proteins in the body.

**anticodon**

A specialized base triplet on one end of a tRNA molecule that recognizes a particular complementary codon on a mRNA molecule.

**antigen**

A non-self protein that is recognized by the immune system and elicits an immune response.

**artificial selection**

The selective breeding of domesticated plants and animals to encourage the occurrence of desirable traits.

**asexual reproduction**

A type of reproduction involving only one parent that produces genetically identical offspring.

**autosome**

A chromosome that is not directly involved in determining sex, as opposed to the sex chromosomes.

**autotroph**

An organism that obtains organic food molecules without eating other organisms. Autotrophs use energy from the sun or from the oxidation of inorganic substances to make organic molecules from inorganic ones.

**backcross**

A genetic cross that reveals whether an individual with a 'dominant' phenotype is heterozygous or homozygous dominant.

**basal metabolic rate**

The minimum number of kilocalories a resting animal requires to fuel itself for a given time.

**biodiversity**

An expression of the number of different species living in a given ecosystem.

**biomass**

The dry weight of organic matter comprising a group of organisms in a particular habitat.

**biological control**

The control of a pest by using a control agent, a predator that feeds on the pest (its prey).

**biotechnology**

The industrial use of living organisms or their components to improve human health and food production.

**carbon cycle**

The movement of carbon between living organisms and the non-living environment, and involving the processes of photosynthesis, respiration and decomposition.

**carcinogen**

A chemical agent that causes cancer.

**carnivore**

An animal that eats other animals.

**carotenoids**

Accessory pigments in the chloroplasts that absorb wavelengths of light that chlorophyll cannot.

**carrier**

An individual who is heterozygous at a given genetic locus, with one normal allele and one potentially harmful recessive allele. The heterozygote is phenotypically normal for the character determined by the gene but can pass on the harmful allele to the offspring.

**carrying capacity**

The maximum population size that can be supported by the available resources.

**central nervous system**

Part of the nervous system (brain and spinal cord) that connects sensory and motor neurones.

**chemiosmosis**

The production of ATP using the energy of hydrogen ion gradients across membranes to phosphorylate ADP.

**chlorophyll**

The green pigment located within the chloroplasts of plants.

**chromosome**

A threadlike, gene-carrying structure found in the nucleus.

**clone**

A population of genetically identical individuals or cells.

**codon**

The basic unit of the genetic code; a three-nucleotide sequence of DNA or mRNA that specifies a particular amino acid or termination signal.

**colony** (of bacteria)

A cluster of bacterial cells (clone) which arose from a single bacterium.

**community**

All the organisms that inhabit a particular area; populations of different species that interact.

**compensation point**

The point at which the rates of photosynthesis and respiration are equal (balanced) in a plant.

**competition**

The struggle between organisms to obtain resources. Competition can be interspecific (between species) or intraspecific (within species).

**conservation**

The management of habitats to maintain or restore species diversity and ecosystem function.

**countercurrent exchange**

The opposite flow of adjacent fluids that maximizes transfer rates; blood in the gills of a fish flows in the opposite direction in which water passes over the gills, maximizing oxygen uptake and carbon dioxide loss.

**crossing over**

The reciprocal exchange of genetic material between non-sister chromatids during synapsis of meiosis I.

**cytochrome**

An iron containing protein, a component of electron transport chains in chloroplasts and mitochondria.

**cytoplasm**

The entire contents of the cell, excluding the nucleus, and bounded by the plasma membrane.

**decomposer**

Organisms, fungi and bacteria that break down dead organic matter to obtain nutrients.

**deforestation**

The removal of trees from a forest, for timber and/or to obtain land for cultivation.

**depolarization**

An electrical state in an excitable cell, whereby the inside of the cell is made less negative relative to the outside than at the resting membrane potential.

**detritus**

Dead organic matter.

**diffusion**

The passive movement of a substance down a concentration gradient from a region of high concentration to a region of low concentration.

**diploid** (cell)

A cell containing two sets of chromosomes (2*n*), one set inherited from each parent.

**dominant allele**

In a heterozygote, the allele that is fully expressed in the phenotype.

**double fertilization**

A mechanism of fertilization in flowering plants (Angiosperms) in which two male nuclei unite with two nuclei in the embryo sac to form the zygote and the endosperm

**ecological efficiency**

The ratio of net productivity at one trophic level to net productivity at the next lower level.

**ecosystem**

A level of ecological study that includes all the organisms in a given area as well as the abiotic factors with which they interact.

**electron transport chain**

A sequence of electron-carrier molecules that shuttle electrons during the redox reactions that release energy used to make ATP.

**endangered species**

A species that is in danger of extinction throughout all or a significant portion of its range.

**enzymes**

A class of proteins acting as catalysts, chemical agents that change the rates of chemical reactions without being used up by the reaction.

**eukaryotic cell**

A cell with a membrane-bound nucleus and membrane-bound organelles.

**eutrophication**

The artificial 'enrichment' of aquatic habitats (streams, rivers, ponds, lakes) by excess nutrients (e.g. due to run-off of fertilizers from farms), resulting in a fall in the oxygen level of the water.

**evolution**

All the changes that have transformed life on Earth from its earliest beginnings to the diversity that exists today.

**extinction**

The process by which a species ceases to exist on Earth, e.g. due to a failure to adapt successfully to a changing environment.

**facultative anaerobe**

An organism that makes ATP by aerobic respiration if oxygen is present but can switch to fermentation under anaerobic conditions.

**feedback inhibition**

A method of metabolic control in which the end product of a metabolic pathway acts as an inhibitor of an enzyme within that pathway.

**fermentation**

A process that makes a limited amount of ATP from glucose without an electron transport chain and produces alcohol or lactic acid as an end product.

**fertilization**

The fusion of male and female gametes to produce a diploid zygote.

**food chain**

The relationship of organisms in each successive trophic (feeding) level in a community.

**food web**

Interconnecting food chains in a community.

**founder effect**

A cause of genetic drift attributable to colonization by a limited number of individuals from a parent population.

**gamete**

A sex cell containing half the number of chromosomes (haploid) as body cells.

**gene**

A discrete unit of hereditary information consisting of a specific nucleotide sequence in DNA.

**gene pool**

The total of all the genes in a population at any one time.

**gene probe**

A short sequence of DNA used as a 'genetic marker'.

**genetic drift**

Changes in a gene pool of a small population due to chance.

**genetic engineering**

The modification of an organism's DNA, usually by the insertion of additional DNA. The changed organism will then have desirable traits (e.g. resistance to disease) which is inherited by its offspring.

**genetically modified organism**

An organism (e.g. a crop plant) which has had its DNA changed by genetic engineering.

**genome**

The complete complement of an organism's genes.

**genotype**

The genetic make-up of an organism.

**global warming**

The increase in the Earth's temperature due to the build-up of heat trapping 'greenhouse gases' (e.g. carbon dioxide) in the atmosphere.

**gross primary productivity**

The total primary productivity of an ecosystem.

**herbivore**

A heterotrophic animal that eats plants.

**heterozygous**

Having one dominant and one recessive allele for a given characteristic.

**homeostasis**

The maintenance by organisms of a constant internal environment.

**homologous chromosomes**

Chromosome pairs of the same length that possess genes for the same characters at corresponding loci. One homologous chromosome is inherited from the father, the other from the mother.

**homozygous**

Having two identical alleles for a given characteristic.

**Human Genome Project**

An international collaborative effort to map and sequence the DNA of the entire human genome.

**immunity**

The ability of an organism to resist disease.

**incomplete dominance**

A type of inheritance in which the F1 hybrids have an appearance that is intermediate between the phenotypes of the parents.

**induced fit**

The change in shape of the active site of an enzyme so that it binds more snugly to the substrate, induced by the entry of the substrate.

**insulin**

A hormone that causes the body to convert excess glucose into glycogen for storage in the liver.

**interstitial fluid**

Also known as tissue fluid, this forms the internal environment of vertebrates, consisting of the fluid filling the spaces between cells.

**in vitro fertilization**

Fertilization of the ova in laboratory containers followed by artificial implantation of the early embryo into the mother's uterus.

**isotonic solutions**

Solutions of equal solute concentration.

**Krebs cycle**

A chemical cycle that completes the metabolic breakdown of glucose molecules to carbon dioxide with the release of energy.

**law of limiting factors**

The rate of a process involving several factors will be limited by that factor which is in shortest supply.

**lignin**

A complex aromatic compound which impregnates the cellulose matrix of plant cell walls, making the wall strong and rigid and impervious to gases, water and solutes.

**linked genes**

Genes that are located on the same chromosomes.

**locus**

A particular place along the length of a certain chromosome where a given gene is located.

**lymph**

A colourless fluid, derived from tissue fluid, found in the lymphatic system of vertebrates.

**lymphocyte**

A white blood cell. The lymphocytes that complete their development in the bone marrow are called B cells, and those that mature in the thymus gland are called T cells.

**lytic cycle**

A type of viral replication cycle resulting in the release of new phages by death or lysis of the host cell.

**meiosis**

A two-stage type of cell division in sexually reproducing organisms that results in gametes with half the chromosome number of the original cell.

**memory cell**

A clone of long-lived lymphocytes, formed during the primary immune response, that remain in a lymph node until activated by exposure to the same antigen that triggered its formation.

**metabolism**

All the organism's chemical processes, consisting of anabolic and catabolic pathways.

**micropropagation**

A technique by which individual plants are grown from a cluster of rapidly dividing cells on a nutrient medium, i.e. in tissue culture.

**mitosis**

A type of cell division that conserves the chromosome number (diploid) by equally allocating replicated chromosomes to each of the daughter nuclei.

**monoculture**

The cultivation over a large area of a single crop plant.

**mutagen**

A chemical or physical agent that interacts with DNA and causes a mutation.

**mutation**

A permanent change in the DNA or the chromosome number of the cell.

**natural selection**

The differential success in the reproduction of different phenotypes resulting from the interaction of organisms with their environment. It is a process that encourages the transmission of favourable alleles and hinders the transmission of unfavourable ones so contributing to evolution.

**negative feedback**

A primary mechanism of homeostasis. When there is a change in a monitored physiological variable a response is triggered to counteract the initial fluctuation.

**net primary productivity**

The gross primary productivity minus the energy used in respiration. It represents the energy in an ecosystem available to consumers.

**neurotransmitter**

A chemical messenger released from the synaptic terminal of a neuron at a chemical synapse that diffuses across the synaptic cleft and binds to and stimulates the presynaptic cell.

**niche**

An organism's role in an ecosystem.

**nucleus**

The chromosome-containing organelle of a eukaryotic cell.

**omnivore**

A heterotrophic animal that consumes both meat and plant material.

**osmoregulation**

The control the water balance in organisms.

**osmosis**

The diffusion of water molecules across a selectively permeable membrane from a dilute solution to a concentrated solution; water diffuses from a region of high water potential to a region of lower water potential.

**oxidative phosphorylation**

The production of ATP using energy derived from the redox reactions of an electron transport chain.

**oxygen debt**

The amount of additional energy required to break down lactic acid, which builds up during anaerobic respiration.

**parasite**

An organism that obtains nutrients from another living organism, known as the host.

**pathogen**

A disease-causing microbe.

**pesticide**

A manufactured chemical used to kill pests in managed environments.

**phage**

A virus that infects bacteria.

**phenotype**

The expression of an organism's genotype; the physical and physiological characteristics of an organism.

**photophosphorylation**

The process of generating ATP from ADP and phosphate by means of a proton-motive force generated by the thylakoid membrane of the chloroplast during the light reactions of photosynthesis.

**photosynthesis**

The conversion of light energy to chemical energy stored in glucose; the process in green plants by which carbon dioxide and water combine, using light energy, to form glucose and water.

**plankton**

Microscopic organisms that occur near the surface of oceans, ponds and lakes.

**plasmid**

A small circular piece of DNA found in bacterial cells.

**plasmolysis**

A phenomenon occurring in plant cells when placed in a hypertonic solution, resulting in the withdrawal of the cytoplasm and the plasma membrane from the cell wall.

**pollination**

The transfer of pollen onto the stigma of a carpel by wind or animal carriers.

**polymerase chain reaction**

A technique for increasing the quantity of DNA *in vitro*.

**population**

A group of individuals of one species occupying a given unit of space at a given time.

**predator**

Animals specialized in feeding on another species (prey).

**primary productivity**

The rate at which light energy is converted to chemical energy of organic compounds by autotrophs in an ecosystem.

**protein synthesis**

The assembly of proteins from amino acid 'building blocks' in the cytoplasm of cells, under the direction of DNA in the nucleus.

**proton-motive force**

The potential energy stored in the form of an electrochemical gradient, generated by the pumping of hydrogen ions across biological membranes during chemiosmosis.

**proton pump**

The active transport mechanism in cell membranes that uses ATP to force hydrogen ions out of a cell and, in the process, generates a membrane potential.

**pyramid of biomass**

A diagram used to represent the decrease in biomass of organisms at each trophic level in a food chain.

**pyramid of energy**

A diagram which represents the total energy requirement of each successive trophic level in a food chain.

**pyramid of numbers**

A diagram used to represent the relative numbers of individuals at each trophic level in a food chain.

**recessive allele**

The form of a gene which, when inherited, is only expressed in an organism's genotype if no dominant allele is present.

**recombinant DNA**

An organism's DNA to which a section of DNA from another species has been added in genetic engineering.

**refractory period**

The short time immediately after an action potential in which the neuron cannot respond to another stimulus, due to an increase in potassium permeability.

**resting potential**

The membrane potential characteristic of a non-conducting, excitable cell, with the inside of the cell more negative than the outside.

**rubisco**

Ribulose carboxylase, the enzyme that catalyses the first step of the Calvin cycle.

**selective permeability**

A property of biological membranes which allows some substances to cross.

**sex chromosomes**

The pair of chromosomes responsible for determining the sex of an individual.

**sex-linked gene**

A gene located on a sex chromosome.

**sexual reproduction**

Reproduction involving two parents, each of which provides a gamete which fuse during fertilization, resulting in offspring that have unique combinations of genes.

**sister chromatids**

Replicated forms of a chromosome joined together by the centromere and eventually separated during cell division.

**sodium–potassium pump**

A special transport protein in the plasma membrane of animal cells that transports sodium out of and potassium into the cell against their concentration gradients.

**speciation**

The origin of a new species in evolution.

**species**

A group of populations of organisms capable of interbreeding with one another.

**species diversity**

The number and relative abundance of species in a community.

**substrate level phosphorylation**

The formation of ATP by directly transferring a phosphate group to ADP from an intermediate substrate in catabolism.

**succession**

A progressive sequence of changes in the flora and fauna of a region, from pioneer to climax communities.

**taxonomy**

The branch of biology concerned with naming and classifying the various forms of life.

**testcross**

The breeding of an organism of unknown genotype with a homozygous recessive individual to determine the unknown genotype.

**translocation**

The transport of sugars and other substances through phloem cells in plants.

**transpiration**

The evaporation of water from leaves, resulting in the movement of water up through the xylem vessels (the transpiration stream).

**trophic level**

A feeding level in a food chain or food web.

**tropism**

A growth response that results in the curvature of whole plant organs toward or away from stimuli that come from one direction.

**vaccine**

A harmless variant or derivative of a pathogen that stimulates a host's immune system to produce a defence response against the pathogen.

**variation**

The differences in characteristics of members of the same species.

**water potential**

The physical property predicting the direction in which water will flow; the tendency for water to leave a system.

**zygote**

The diploid product of the fusion of haploid gametes in sexual reproduction.

# Further reading

**General Text Books**

*Biology* (1992), Rowland, University of Bath Science 16–19, Nelson Thornes (ISBN 0 17 438425 4)

*Biology: Principles and Processes* (1993), Roberts, Reiss & Monger, Nelson Thornes (ISBN 0 17 448176 4)

*A-level Biology* (reprinted 1996), Phillips & Chilton, Oxford Univ. Press (ISBN 0 199 14584 9)

*Understanding Biology* (3rd edn 1995), Toole & Toole, Stanley Thornes (ISBN 0 7487 1718 8)

*Biological Science* (3rd edn 1997), Green, Stout & Taylor, Cambridge Univ. Press (ISBN 0 521 56178 7)

*Advanced Biology* (2000), Kent, Oxford Univ. Press (ISBN 0 19 914195 9)

*Advanced Biology* (1997), Jones & Jones, Cambridge Univ. Press (ISBN 0 521 48473 1)

*Central Concepts in Biology* (1995), Jones *et al.*, Cambridge Univ. Press (ISBN 0 521 48501 0)

**Reference Books**

*Advanced Biology Topics: Biodiversity – The Abundance of Life* (1997), Chapman & Roberts, Cambridge Univ. Press (ISBN 0 521 57794 2)

*Mammals: Structure and Function* (1998), Clegg (Illustrated Advanced Biology Series), Murray (ISBN 0 7195 7551 6)

*Signs, Symbols and Systematics*, the ASE companion to 5–16 Science, The Association For Science Education (ISBN 0 86357 232 4)

*Biological Nomenclature* (3rd edn 2000), ed. Cadogan, Institute of Biology (ISBN 0 900 49036 5)

*Microbiology and Biotechnology* (1995), Cadogan & Hanks, Nelson (ISBN 0 174 48227 2)

*Microorganisms and Biotechnology*, Taylor, University of Bath Science 16–19, Nelson Thornes (ISBN 0 333 48320 0)

*Microorganisms, Biotechnology and Disease* (1991), Lowrie & Wells, Cambridge Univ. Press (ISBN 0 521 38746 9)

*Microbiology and Biotechnology* (1994), University of Cambridge Local Examinations Syndicate, Cambridge Univ. Press (ISBN 0 521 42204 3)

*Applied Genetics* (1991), Hayward, University of Bath Science 16–19, Nelson Thornes (ISBN 0 333 46659 4)

*Applied Ecology* (1992), Hayward, University of Bath Science 16–19, Nelson Thornes (ISBN 0 174 48187 X)

*Genetics and Evolution* (1999), Clegg (Illustrated Advanced Biology Series), Murray (ISBN 0 7195 7552 4)

*Cell Biology and Genetics* (1996), Adds *et al.*, Nelson (ISBN 0 174 48266 3)

*Revision Express AS Fast-Track Biology* (2001), Rowlands, Pearson (ISBN 0 582 43233 2)

This is not a prescriptive list and any further reading should be carried out after consultation with your teacher.

# Index